The Thought of Matter

Philosophical Projections

Series editor: Andrew Benjamin, Distinguished Professor of Philosophy and the Humanities, Kingston University, UK, and Professor of Philosophy and Jewish Thought, Monash University, Australia

Philosophical Projections represents the future of Modern European Philosophy. The series seeks to innovate by grounding the future in the work of the present, opening up the philosophical and allowing it to renew itself, while interrogating the continuity of the philosophical after the critique of metaphysics.

Titles in the Series

Foundations of the Everyday: Shock, Deferral, Repetition, Eran Dorfman
The Thought of Matter: Materialism, Conceptuality and the Transcendence of Immanence, Richard A. Lee
Deception, Violence and Law: Renewing the Political, Peg Birmingham

The Thought of Matter

Materialism, Conceptuality and the Transcendence of Immanence

Richard A. Lee Jr.

ROWMAN &
LITTLEFIELD
INTERNATIONAL
London • New York

Published by Rowman & Littlefield International, Ltd.
Unit A, Whitacre Mews, 26-34 Stannary Street, London SE11 4AB
www.rowmaninternational.com

Rowman & Littlefield International, Ltd. is an affiliate of Rowman & Littlefield
4501 Forbes Boulevard, Suite 200, Lanham, Maryland 20706, USA
With additional offices in Boulder, New York, Toronto (Canada), and London (UK)
www.rowman.com

British Library Cataloguing in Publication Information Available
A catalogue record for this book is available from the British Library

ISBN: HB 978-1-7834-8642-7
ISBN: PB 978-1-7834-8643-4

Library of Congress Cataloging-in-Publication Data

Lee, Richard A., Jr.
The thought of matter : materialism, conceptuality, and the transcendence of immanence / Richard A. Lee Jr.
pages cm. -- (Philosophical projections)
Includes bibliographical references and index.
ISBN 978-1-78348-642-7 (cloth : alk. paper) -- ISBN 978-1-78348-643-4 (pbk. : alk. paper) -- ISBN 978-1-78348-644-1 (electronic)
1. Materialism. 2. Matter. I. Title.
B825.L46 2015
146'.3--dc23
2015031456

∞™ The paper used in this publication meets the minimum requirements of American National Standard for Information Sciences Permanence of Paper for Printed Library Materials, ANSI/NISO Z39.48-1992.

Printed in the United States of America

Dedication

In memoriam Sam IJsseling (1932-2015)
No philosopher has taught me more about the transcendence of materiality.
From now on, every day is lived in after the end of the world.

For the journey-level pipefitters and apprentices of Local 597 of the United Association of the Plumbing and Pipefitting Industry of the United States and Canada.

Contents

Acknowledgements

I would have completely given up on philosophy if it were not for the fact that students seem to think I might have something to say. I have benefited in ways that are immeasurable from a Graduate Seminar I taught at DePaul University on Adorno's *Negative Dialectics*. The comments, critiques, and contributions of the participants were daunting, exciting, thoughtful, and serious. My reading of Adorno stands on the shoulders of the participants in that seminar. I am also grateful to the undergraduate students in my Capstone Seminar who so graciously listened to my working out of this book. Whatever I say clearly is a direct result of their intervention. I could not have asked for a more thoughtful and wickedly funny group of interlocutors.

Without their knowing it, three of my colleagues—I call the group "Doctors who Drink Together"—have also provided me with the opportunity to test out my ideas and to be challenged. Profs. Will McNeill and Sean Kirkland, both in the Department of Philosophy at DePaul University, and Prof. Liam Heneghan of the Environmental Science and Studies Department, have inserted themselves into the core of my thought in ways that I will never be completely aware. Their arguments, taunts, jibes, cajolings, and shared commitment to what "matters" have pushed this project along.

Two people have been journey-level workers in that they have read a good deal of the manuscript and provided remarkable encouragement, often in the form of challenges. Sean Kirkland is a remarkable philosopher, no matter the topic. He has been as willing to read my interpretation of Aristotle, his area of expertise, as my discussions of Marx and more recent Marxists, not his area of expertise. But he is a philosopher whose instincts are always just so. He has helped, altered, and, at a certain moment, almost destroyed this project in ways that I so appreciate. Prof. Jennifer Hansen, of St. Lawrence University, has been a similar journey-level worker. Much more out of her expertise, her insights, encouragement, and close reading have made this book better than it could have been. I thank both of them and, as they are well aware, many drinks will be purchased for their help!

I am lucky to work in a department that is populated by philosophers of the highest order. I am thankful to all of them for their friendship, support, and philosophical conversation. I am sure that I have cribbed many of their insights and for that I apologize and hope this acknowledgement goes some way toward stemming a lawsuit. But among a department of fantastic colleagues, Michael Naas, the Chair of the Department, stands out. For us as a department and for me as an individual, he has, certainly, a claim on our messianic powers. I am thankful for his friendship, his support, and, perhaps above all, his serving as a constant reminder of what it means to do philosophy. Andre Agassi once said that the difficulty with playing against Pete Sampras is that, already in the locker room, you knew that you could play the best tennis of which you are capable and still lose. But knowing that forces you to play better tennis. While I may be no Andre Agassi, Michael is certainly Pete Sampras!

As the Introduction and final chapter make clear, my thought in this book has been fundamentally positioned by my aunt Betty and my godmother, Aunt Peggy. If materialism means anything worth thinking, it must engage the ways in which they continue to make a claim on my weak messianic power.

Introduction

I did not set out to write a book about the philosophical approach to matter known as materialism. My initial aim was a bit more modest: I wanted to write a book about Hobbes, primarily focused on his text *De Corpore*,[1] a text that I took, and still take, to be an attempt at a materialist philosophy, that is a philosophy that—in one way or another—sets out from the reality of material things in order to insist on their priority over thought. Yet I came upon a methodological problem in Hobbes that perplexed me to the point that I could not continue. I can put the problem in several ways. First, starting from *De Corpore* itself, it is clear that, for Hobbes, only those things are, exist, that are matter and that matter is always in motion. Early on, Hobbes already took the term "to be" as a synonym for "body," that is, an extended thing that is material.[2] While his ontology acknowledges only such material things, the *De Corpore* begins with a discussion of method, the proper method of philosophy. That method begins with the correct understanding of names, moves on to propositions and syllogism, and ultimately concludes with a discussion of composition and resolution, that is, how to take what is given (those material things his ontology insists are the only things that exist) and understand it in terms of its causes. These causes, in turn, must not be counted as existing things because they are not material, nor always reducible to material things, and they are certainly not given in our sensory experience of material things. This left me with an interpretive dilemma: either I say that Hobbes is caught in a circle in that he makes a claim (whatever exists is material) that cannot be made on its own terms or Hobbes is not a materialist.

Stepping aside from the peculiarities of that text, the problem can be put more broadly: the metaphysical claim that existence is identical to being material cannot itself be grounded materially unless one is to assume that which is to be demonstrated. Yet one often finds, either implicitly or explicitly, an understanding of materialism as the philosophy that claims "whatever exists is a material thing." This form of the claim, however, shares certain features with

Hesiod's *Theogony*: "first came Chaos." That is, it seems to be the furthest thing from a philosophical position. Matter may be, in this case, the ground of everything philosophically subsequent to its positing, but it, therefore, must simply emerge or be posited. Thus matter is more akin to a divinity than to a material thing.

A third way to put the problem draws out the methodological difficulties: philosophy is a way of thinking about something and yet matter is not just the thought of it but the *very thing thought*. There is, therefore, an irreducible difference between thought and matter that will always trouble the philosophy that attempts to think matter as prior. In short, materialism seems impossible. It became clear to me that I had to somehow deal with this problem in order to present an adequate interpretation of Hobbes's *De Corpore*. The present book is my attempt to formulate, as clearly as possible, the difficulty and think through a few paths that have been taken to think matter and provide a materialist philosophy that takes this difficulty seriously.

Some clarifications are in order.

First clarification: while I develop the reasons for this, I have come to define matter in a way that might not seem intuitive. Matter is that which is other than thought. I have come to this definition because I struggled to see just what it is that is characteristic of something that comes to be called "matter." Some of the likely candidates, for example, extension, turn out not to be acceptable precisely because these candidates are not themselves material. Through Aristotle and Adorno, I have come to see the peculiar feature of matter is that it is not what thought is. This feature, being other than thought, is what I came to call "materiality." There can be, therefore, discussions of matter that, either explicitly or implicitly, turn it into the thought of it, recognizing only those aspects of material things that are thinkable. These approaches are not, therefore, materialist. However, there is a danger on the other side. There are many philosophical approaches that not only simply posit matter like some sort of chthonic divinity, but also posit some way in which we have access to that matter outside of thought. To be sure, I do not want to deny that we do encounter, engage, and experience matter in a wide variety of ways: sensation, digestion, affection, physical interaction. The problem for me is that, as Hobbes put it, sensation is not philosophy, even if philosophy begins in sensation. The same would hold true of all the other varieties of engagement and encounter. There is something crucial to the philosophical approach to matter to which I want to cling: it can provide a critical analysis in ways that sensation, digestion, affection, and so on, do not. This is something I learned from Aristotle's *Nichomachean Ethics*. I can certainly eat a chicken. However, the act of eating does not always allow me to understand whether it is good (in whatever way) to eat a chicken. Thought, and therefore philosophy, is crucial as a way to understand the operation of matter and that understanding is crucial to a critical engagement with our world. Therefore, I will call a philosophy "materialist" if it not only prioritizes the otherness of matter to thought but also thinks the relation of thought to matter so as to allow a critical engagement.

Second clarification: the philosophical discussion of matter is an old one. Because of its antiquity, there are many more philosophers whose discussions of matter merit attention than can be reasonably engaged in any single text, perhaps in any single lifetime. What is more, not every philosophically important discussion of matter can reasonably be called materialist as not every discussion of matter takes it to be ontologically fundamental. Descartes, for example, has a robust, important, and interesting discussion of matter and yet is not a materialist precisely because he takes thought to be rooted in an existing thing (a substance) that is, by definition, immaterial, namely the soul, the "thing that thinks." What is more, he even demonstrates that I must begin with thinking as thought is more available to me than an object that seems to be outside of thought. What a thing is, for Descartes, is available to thought in such a way that its "reality," its "being" is not relevantly different to what is already encountered in the thought of it. He never, therefore, struggles with the basic problem with which I was presented by Hobbes, the problem that if there is something constituted by being material there is, nevertheless, the thought of it that is not identical to it.

I take Descartes here only as an example of a principle I have used in the selection of paths I present here. Over the past ten years (that was when I set out to write on Hobbes!), I have frequently been asked what I am working on and when I respond, "matter and materialism," I am almost always met with the name of a philosopher on whom I *absolutely must* be working: Democritus, Engels, Lucretius, Badiou, Hegel (yes, even Hegel!), Heidegger, Meillassoux, Aquinas, Malibou, and so on. I have enjoyed responding, "No, I am working on the usual materialists like Duns Scotus." The reality is that my obsession was precisely with the question of method—how can we think that which is other than thought? All of the thinkers I have mentioned and many, many more certainly have fundamental things to say about matter but I did not find them helpful for thinking through this basic problem. I could, perhaps, engage each or all of them to show why they are not helpful. However, such a project would not be helpful in trying to uncover and untangle the specific problem of *how* thought can be of something other than it while maintaining precisely that difference.

There is much current discussion of materialism that arises directly or indirectly from an approach that is variously labeled "speculative realism" or "object oriented ontology." It will be clear from even a cursory review of the table of contents that I neither engage nor treat that approach directly in this text. Given its relative importance in contemporary philosophy, I feel a few words are in order about my choice in this regard.

First, anyone familiar with this approach will recognize the many ways in which I circle around and at times enter directly into its field and insights. At times, I am in agreement, at times I disagree strongly. I do not pause to point out these moments of agreement and disagreement. I urge those interested to take up Meillassoux, Graham Harman, and others working in this tradition. I have been helped in many ways by the work of Adrian Johnston to understand when and how my path crosses over the path of speculative realism. There is one basic insight Meillassoux provides in *After Finitude* that I think is crucial to any materialist philosophy: objects exist outside of their givenness. Translated slightly, in a way I do not think he would reject, there is something that is other than thought whose existence is not exhausted by what is given in thought. This "existence outside givenness" is what I call "materiality." Yet there is something that nags me at this point that returns me back to my basic methodological problem, namely that this no philosopher has ever denied, perhaps with the exception of Berkeley. So in a sense, my project sets out to intervene at precisely this moment. Given that there are objects whose existence is not exhausted by their givenness to (or appropriation by) thought, how can we redeem that existence *when they are thought*, as they are, for example, in Meillassoux's own philosophy? That "how" is for me always a question of the medium. These material things, Meillassoux's "ancestral objects," have a mode of existing that is not the same as their mode of existing in thought, yet they are thought, even by the speculative realist. Thought is a medium through which they are given even when one claims that they exist outside of givenness. There is, for lack of a better term, mediation. No insistence on a mode of being outside of mediation will ever prevent these objects from mediation. They are thought and thought is a medium through which even their nonmediated existence is mediated.

I apologize for the convoluted way of expressing the problem, but I do not see a way around it. When Meillassoux explains that what he is interested in uncovering is how to make sense of the claims of science about objects that exist outside of givenness, he already engages a medium, namely science, through which the non-given is, in fact, given. Let me be clear: I am not trying to outplay speculative realism by some logical or methodological trick. Rather, I am trying to see how the speculative realist raises but does not clarify my problem. Science is a way in which even the arche-fossil (that which exists before or outside of givenness) is given and I find the claim that philosophy must make good on the claims of science in relation to the "ancestral" to be in need of critical engagement. Yet this, perhaps, shows more agreement than disagreement if I amend the speculative realist claim in a way that those in that tradition might not accept. Namely, what I call "materiality" is precisely that way of being that is other than thought but that way of being is and must be thought. Therefore, the "how" is crucial and I do not find thorough or acceptable answers coming from the speculative realist approach. I am a fellow-traveler in many respects but perhaps not in one that the speculative realist would find essential to that approach.

Third clarification: while much of the material presented here deals with figures from the history of philosophy (Aristotle, Duns Scotus, Hobbes, Marx), it is not a history of materialism. If this is not clear from the progression of chapters (Aristotle, then Althusser, Marx, Adorno, then Scotus, etc.), I hope it is clear from the methodological discussion above. I begin with Aristotle not because he is historically first but because I find his thought insightful for getting clarity on the "trouble" that matter causes for philosophy, what I often refer to as its "errancy." Chapter 1, therefore, is about Aristotle because in his philosophy we see the exposure of matter as other than thought *and* the recuperation of that otherness for thought. This is not meant as a criticism of Aristotle. It is meant to show the basic problem, in fact that very basic-ness of the problem of *how* to think that which is other than thought. That is why, in chapter 2, I turn to Althusser, Marx, and Adorno; for having seen the basic problem, they provide an insight to its analysis. That analysis, however, requires an expansion of what I think of as our "ontological beastiary." Frequently, we tend to think the terms "real" and "existing" either as synonyms or at least as coextensional. Marx gives a powerful argument about why that is troubling, but Duns Scotus, on whom chapter 3 is focused, provides the ontology adequate to thinking that which is other than thought by showing that there must be things that are "real" that, nevertheless, do not actually exist. His insistence on the reality of what he calls "common natures" that does not entail their "existence" leads him to the remarkable conclusion that materiality is simply singularity. Singularity, the irreducibility of difference, stands outside the grasp of understanding, outside of thought. This helps me understand my original difficulty with Hobbes, to whom I turn in chapter 4. Having this ontological expansion on the table, however, helps to understand another feature of Hobbesian materialism, a feature that is central to Marx, Althusser, Benjamin, and Adorno, namely its social basis. This allows me to turn, in chapter 5, back to Adorno and Benjamin to see what this renewed understanding of materialism can provide for philosophy and why philosophy must be *redemptive*.

This term, "redemption," may seem out of place in a discussion of materialism. Yet both Benjamin and Adorno insist on its centrality. This centrality helps to understand why I argue the importance of not rejecting thought in favor of some immediate approach or access to matter or the material thing. That is, I see an intimate connection between the task of redemption and the promise of thought, a thought that proceeds in the light of the redemption of that which is other than thought. A few years ago, I was at the funeral of my aunt Betty—a wonderful woman whose generosity, warmth, and laughter were at odds with her brutal, abusive, and horrific life at the hands of her husband. My godmother and Betty's sister-in-law, Aunt Peggy, was at the funeral. She took me by the hand to (re)introduce me to relatives, friends, the priest, and other guests. Her introduction was always the same, "This is my godson, Rick, he's a professor . . . OF PHILOSOPHY!" One thing was immediately clear to me: Aunt Peggy was proud of me and my profession. Her husband, my godfather, Uncle Dan, was a pipefitter, as were and are a great many of the men in my family. While I did not ask, I have the suspicion that she really had no idea of what I do. Certainly she has never read any of my work. For all I know, she is a speculative realist! She knows I teach. She knows I am a professor. Certainly one can understand the pride a relative would take in my achieving a certain rank and status. What I could not understand at the time, however, was her insistence on the "philosophy" as part of her pride. What is the origin of this pride?

This moment is one upon which I have thought over and over and it suddenly opened up for me Benjamin's insight—an insight he calls "historical materialism"—that the past makes a claim on us and, therefore, we have a power to make good on that claim.[3] Aunt Peggy made me see what Benjamin means when he says "our existence on earth was awaited." There is in the very fact of living, of "going on," a projection of or openness to the future, to subsequent generations. Even merely going on is a hope. But hope always entails an "otherwise," because no one hopes for what is. The seeming emptiness of this hope—"otherwise"—is, rather, a kind of infinity. The hope for something other is a hope that shoots out in an infinite number of directions, like rays of the sun, only one of which is my being a professor . . . OF PHILOSOPHY.

Yet if this hope is like a ray from the sun, it can be traced back in the other direction. A ray is, after all, a ray *of the sun*, and this hope, is, after all, a hope of the past. When I trace this hope back, I trace it back to its infinity, that is, to the indeterminate "otherwise." And when I realize that my presence and my being a professor of philosophy is but one point on one of those rays, I now encounter something like a responsibility. That I am at all and that I am a professor and that I am a professor OF PHILOSOPHY is but one otherwise. But the hope was open, if not infinite. Therefore, tracing the hope back I am now responsible to its infinity and, as a result, I do not "automatically" redeem that hope just by being.

Returning back to the pride, which I have now thought is bound up with hope, my attempt to make good (a literal parse of "redeem") on that pride must also make good on the "philosophy" that was part of that hope. How can Aunt Peggy hope for that which she does no know? There is something implicit in philosophy, expressed in her hope and pride, that has a redemptive power. This places on me a responsibility—Benjamin calls it a "weak messianic power"—to make good on the promise of philosophy. My interest in materialism is, I would argue, a necessary moment in that redemptive attempt. Pipefitters certainly do engage matter in all its materiality. Aunt Betty encountered matter every time her husband brutally beat her. But Aunt Betty's generosity and laughter was a recognition too of an "otherwise." What Benjamin calls our "weak messianic power" must be, I now see, a way of recognizing that I cannot make good on that hope by giving up on the critical power of thought.

For these reasons, as I hope chapter 5 makes clear, I cannot give up on thought. Here I stand, as I argue in this text, with Marx,

Althusser, Adorno, and Benjamin. We do not think matter by saying to Aunt Betty, "well, this is all there is." This kind of immediacy she knew all too well. But every tray of mostaccioli, every kielbasa, every joke, projected out an otherwise that is now a condition of my being. And the power to make good on that is "weak" because this thought of materialism is necessary, but all too far from sufficient, to redeem it.

NOTES

1. Thomas Hobbes, *De Corpore*, ed. Karl Schuhmann (Paris: J. Vrin, 1999).
2. Thomas Hobbes, *Critique Du De Mundo de Thomas White*, ed. Jean Jacquot and Harold Whitmore Jones (Paris: Librairie Philosophique J. Vrin, 1973).
3. See chapter 5 for a fuller discussion of this claim.

ONE

Aristotle's Critique of Thales

There is a story that Thales is the first philosopher in the West. And what is philosophical about what we know of him? He is said to have said that all things are from water. We have this on good authority, that of Aristotle. It is not that Aristotle is faithful in his interpretation of Thales. Indeed, if it were not for Aristotle—and Plato before him—we would not have any Thales about which to think. The interpretation of Thales, therefore, is at the same time the "text" of Thales. For Aristotle, Thales is the first philosopher because he is pursuing the question of principle (ἀρχή). Aristotle begins *Metaphysics* not by raising the topic of being, of what is, or even of οὐσία. Rather, he begins with the claim that everyone desires to "know" [εἰδέναι]. As the discussion unfolds, this issue of knowledge turns toward an "of"—everyone desires to know what? Why begin with the question of grasping what is in thought before addressing the central issue of the question of what is? This must be because no other form of knowledge, for example intuition (νοῦς) or practical knowing (φρόνησις), is appropriate to understand that which is without qualification. Therefore, to approach what is just as it is must require a peculiar form of knowing. This is not immediately obvious, as the text indicates. One might think that a certain skill with things [τέχνη] is better at getting at what is, or even a more direct grasp [ἐμπερία] would be better.[1] After seeing their relative merits, Aristotle concludes that wisdom is "knowledge having to do with certain principles and causes."[2]

The question that opens the very next chapter follows directly from this. If wisdom is knowledge of principles and causes, then just what kind of principles and causes are those? In chapter 3, Aristotle notes that "of the first philosophers the majority thought that the causes in the form of matter were alone the principles of all things" (Meta. 983b7–8). For Aristotle, Thales's position is ultimately about knowledge since he claims that there is a principle of all things and that principle is water.[3] As a principle, water makes things intelligible because it is the role of a principle to bring something to intelligibility. The first of those who can be called philosophers sought the intelligibility of things in "that from which all entities come, from which each thing primarily arises and into which it is at the end resolved, the substance remaining but changing as to affections" (Meta. 983B8–12). What makes Thales a philosopher, therefore, is not the claim that all things are water or from water but rather the understanding of all things through a principle. While Thales could be wrong about just what that principle is, he is certainly a philosopher because he is asking after the principle of things.

In this way, Aristotle links the notion of principle with the notion of intelligibility or understanding, in particular that form of understanding he calls wisdom (σοφία). Wisdom is a form of knowing that Aristotle arrives at by means of distinguishing it from other forms of cognitive and gnoseological abilities.[4] Wisdom cannot be any of the senses because "they do not . . . give the reason for anything, e.g. as *why* fire is hot, but merely indicate *that* it is hot" (Meta. 981B12–14). What is more, the senses are the most important forms of knowledge when it comes to particulars,[5] but they can only provide access to that which is immediately given. The senses cannot provide an account (λόγος), that is, understand *why*. Aristotle points to a difference between what we might call the grasp in thought (or, in the case of sensation, even the bodily grasp) of what is given as it is and the understanding of *why* something is or is the way it is. Thales is a philosopher, it would seem, because he turns away from what is immediately given and attempts to understand it in terms of a principle.

Aristotle's concern with Thales begins from Aristotle's assumption that Thales is seeking a principle. Since the grasp of principles is the giving of an account, that is, the attempt to grasp the *why* in the face of the immediate and obvious fact *that* something is, is σοφία, Thales is a pursuer of wisdom, a philosopher. This relies on the recognition that the claim that everything is from water presupposes a difference between what presents itself to sensation (*this* beer, *this* bar, etc.) and *why* it is. This originary dichotomy is, on Aristotle's account, the necessary and perhaps sufficient condition of philosophy and it is his recognition of the original duality of appearing thing and account that makes Thales, according to Aristotle, the first philosopher.[6]

Skill (τέχνη) brings us closer to wisdom because in skill there is a grasp of the causes. Those with an art, a skill, a τέχνη, are thought to be closer to σοφία than those that merely have experience. "And the reason for this is that the skilled know the cause, whereas the experienced do not. For the experienced know the 'that' but not the 'because,' whereas the skilled have a grasp of the 'because,' the cause" (Meta. 981A12–19).[7] Yet what appears here to be a stark distinction is not as strict as it might seem. If we take the example of medicine, we can say that the doctor must treat *this person* and not the human in general. In a field such as medicine (and this can be extended to architecture, painting, and all other forms of τέχνη), the goal is to *do* something and such a practical action always has to regard the singular that is given.[8] The person who acquires the skill, therefore, but is not experienced, and therefore is not knowing particulars, will often appear less wise than the person of experience.

So wisdom is not at all like the senses, because, unlike the senses, wisdom understands the "because." Wisdom shares certain characteristics with τέχνη because, like τέχνη, it grasps the reason why, but, unlike τέχνη it can never appear to be less than wisdom. Craft (τέχνη) is, by its very nature, directed toward action and, therefore, always toward the particular. However, the structure of craft is directed away from what is given, and therefore away from the particular, and therefore can appear to be less wise than experience. The "journeyman," that is, the one who has a τέχνη, knows not just how to make *this* statue, install *this* heating system, but has general knowledge of the causes, of the *why*, such that this knowledge can be applied in particular situations. Wisdom, for its part, will never be directed at the singular, except that it ought to understand these very singulars. Wisdom ought to be the knowledge of the *why* of *this* thing that is.

This raises a tension, and because there is a tension, it raises a difference. There is, on the one hand, the grasp of what is given immediately, and this grasp comes by means of sensation, memory, and experience. This grasp, however, is limited in that it leaves unknown the "because" of whatever is given. One can say that the difference here is between grasping what is immediately given and being able to step outside of what is immediately given in order to understand and, perhaps more importantly, analyze it. Aristotle, therefore, argues that there is a way to step outside of what is given in order to be able to present how and why it is given as it is. He provides, in short, what will come to be called the possibility of the critique of what is given.

On the other hand, the step away from what is immediately given is, for Aristotle, asserted to be a way to *understand* what is immediately given. Here the tension emerges. While Aristotle diagnoses a problem with the immediacy of what is given, in that those ways of grasping it as immediately given are stuck within that immediacy and, therefore, cannot give an account of what is immediately given, the step outside that immediacy is at the same time a move away from it.

> And from these thinkers, then, one might think that there is only a single cause, that said in the form of matter. But as they continued in this way, the facts themselves guided them and forced them to seek further. For even if, as much as you like, all coming to be and destruction are from some simple thing or from several, why does this happen and what is the cause? For indeed it cannot be that the substrate itself forces itself to change. I mean that, for instance, neither wood nor bronze is the cause of either of them changing, neither does wood make a bed or bronze a statue, but there is some other cause of the change. And to seek for this is to seek for the second kind of principle, as we would say, that from which comes the beginning of the change. (Meta. 984A18–25)

Thales was the first philosopher, but he erred in understanding the enterprise itself. Matter, Aristotle insists, can never provide the reasons why generation and destruction occur.[9] If one cannot provide the reasons why something comes to be and passes away, then one is trapped within the givenness of the "it is." This immediacy does not allow a way to understand that which is given to our sensation, memory, and experience. So Thales failed precisely because he was a philosopher. If he had not searched, according to Aristotle, for the *arche*, then he might have been a man of skill and maybe even of experience, but certainly not a philosopher. Only because, on Aristotle's account, he sought after the principle, only because he was, in short, the first philosopher, can he be said to have gone wrong.

The result of Aristotle's critique of Thales is that while Thales was attempting to understand that material thing immediately given, matter as such can never provide the path toward understanding. There is, then, the tension of methodology: in attempting to understand the material thing given to me I need to move away, I am forced by the very attempt of understanding to move away, from the materiality of what is given to me. The distance opened up between the theoretical grasp of material singulars and the matter, the materiality of those singulars themselves also opens a methodological problem for philosophy as Aristotle thinks its task in the *Metaphysics*. As Aristotle shows through his reading of this first philosophical gesture, the very attempt to understand a material singular requires a move away from its materiality, for matter cannot be a principle that gives the *why* or the *because* but only provides the *that* of anything that is, in and of itself requires a step back or, at least, to the side—the path cannot be straight but is necessarily errant. This move away, however, indicates that, for Aristotle, this understanding in the form of σοφία, that is, the understanding that can give an account of why something is at all and is the way it is, necessarily must be other than matter and materiality. Since thought, whether as νοῦς, ἐπιστήμη, or σοφία, requires a turn away from matter, one is forced to say that matter is other than thought, an other *of* thought. Yet at the same time, understanding the *why* sets out from that very material thing that is given, primarily in sensation. So thought is supposed to uncover the principles of that which, by definition, is other than itself. Aristotle argues that wisdom is the grasp of that which is other than thought. The methodological problem, however, is whether in grasping that which is other than itself, thought is able to grasp matter in its otherness, that is, without reducing matter to thought itself.

There is, therefore, a move in Aristotle toward explanation, toward understanding why something is and is what it is.[10] Aristotle argued that this move toward explanation requires the grasp of a principle. By means of a principle, one can come to understand not just *this* that is here before me, that is, I can grasp not only that it is in its givenness to me, but I can grasp *why* it is at all, *why* it is the way it is, and, perhaps most importantly, *what* it is. In the end, what I cannot grasp in immediacy is the mode in which that immediately given appears, that is, is presented to me at all. This means that what presents itself immediately does not also present, in that immediacy, its own account.[11] The distinction Aristotle makes between sensation and practical know-how, on the one hand, and wisdom, on the other, shows what is at stake here for him. Both sensation and practical know-how remain with what is given *as it is given*, and therefore cannot grasp either the way in which it is given nor why it comes to be given at all.

The example of Thales is telling in two complementary directions. First, it is clear to Aristotle that everything that is does not give itself as water. In the face, for example, of wood, fire, a house, or a horse, Thales insists that everything is (from) water. This indicates, for Aristotle, that Thales is not content to remain with the givenness of concrete, singular things as such. Thales turns away from the way in which things show themselves and argues that, despite what is given, there is something else, water, that is necessary if one wants to understand this singular thing given here and now. It is this very move that Aristotle points to as philosophical. All animals have sensation and, as a result, all animals grasp *this* singular here. Many humans have the ability to build houses, heal animals, install heating systems, and so forth, and, as a result, can grasp both *this* singular here and some general conditions and rules regarding it. But beyond even the move toward cause and principle present in *techne*, there is still something left unknown, beyond consideration. Thales does not insist that the principle of all things is water because he wants to build, heal, or install, but rather because he wants to understand why this thing is and is the way it is. What is more, for Aristotle, this gesture means that he is also searching for the mode of the givenness of what presents itself.

On Aristotle's reading, Thales is the first philosopher precisely *because* of his turn away from what is given in sensation. For Aristotle, ultimately, that should also entail a turn away from matter. That is, there is a gesture in the claim "Everything is (from) water" that requires Thales to turn away from a material principle. Given Aristotle's separation of wisdom from skill and even practical know-how, this should also force Thales to turn away from all the material concerns of one's immediate context. Wisdom, Aristotle argues, demands a flight from the material concerns of life precisely because it demands an understanding of the *why*.

Yet it is not as if Thales was reputed, even by Aristotle, to stand outside the world of practical concerns:

> There is the anecdote about of Thales the Milesian and his financial device, which involves a principle of universal application, but is attributed to him on account of his reputation for wisdom. He was reproached for his poverty, which was supposed to show that philosophy was of no use. According to the story, he knew by his skill in the stars while it was yet winter that there would be a great harvest of olives in the coming year; so, having a little

> money, he gave deposits for the use of all the olive-presses in Chios and Miletus, which he hired at a low price because no one bid against him. When the harvest-time came, and many wanted them all at once and of a sudden, he let them out at any rate which he pleased, and made a quantity of money. Thus he showed the world that philosophers can easily be rich if they like, but that their ambition is of another sort. He is supposed to have given a striking proof of his wisdom, but, as I was saying, his device for getting money is of universal application, and is nothing but the creation of a monopoly.[12]

Aristotle repeats here a claim similar to the one in *Metaphysics,* namely that Thales exhibits wisdom, even if he turns out to be wrong. We have here the second direction in which Thales is a telling example for Aristotle. In *Metaphysics,* Thales was labeled the first philosopher because his turn away from the givenness of material singulars was a turn toward principles. In the *Politics,* however, his turn toward principles allows Thales to understand something that is crucial to the material singular. The givenness of material singulars would never allow Thales to understand that there would be a bumper crop of olives. As a philosopher, as someone in search of wisdom, however, he came to understand why olives grow, why the stars have an effect on crops, and thus could bring his wisdom back to the very material singulars his initial turn left behind. In this way, Aristotle shows that this move away from the givenness of what is will always be in the name of understanding what is and its givenness. The financial success of Thales's endeavor means that it is always directed at the concrete singular. Aristotle's reference to the practical, indeed financial, dimension of Thales's wisdom highlights the fact that, although wisdom requires a withdrawal from the material singular, that withdrawal is *in the name of* that material singular.

Furthermore, Aristotle here opens another dimension of the material, one that is no less complex than the first. In this context, Aristotle is arguing that wealth-generation is a responsibility of the *polis,* of the state. By raising this legend of Thales in this context, Aristotle admits that in order to understand something as concrete as population, survival, wealth, and ultimately class and the division of labor, one needs to move away from what is immediately given and turn toward a way of grasping the *why* of that which is given. As such, without a turn away from that immediacy, one would be unable to grasp the very materiality of such complex phenomena as production and wealth-generation.

Still, Aristotle indicates a fundamental problem with matter. Thales's position that everything is or is from water certainly presents water—an elemental form of matter—as a principle of things, namely as that from which things (indeed, all things) come. In *Physics*, Aristotle argues that "if, of things by nature, there are causes or principles of which those things are composed primarily and from which they come to be not accidentally, but come to be what each of them is called according to its *substance*, then everything which is generated is generated from a subject and a form."[13] To this extent, Aristotle seems to agree with Thales and other philosophers who posit a material principle for things in that he insists that the underlying material subject is a crucial factor in the actual being of any given thing. But if Thales is correct in saying that water is the principle of any given thing, then he must be saying that water is the reason why the table is a table. This seems to be a difficult route toward understanding and explanation. If water is to perform its function as a principle of a thing that gives itself as other than water, then Thales must acknowledge that there is a something, for example, a tree, that comes to be from water. Water, for Aristotle, here functions as the principle, and therefore the explanation, of why the tree is. The question now hinges on whether water as water can sufficiently explain the actuality of a tree. He argues that if I start from the actuality given to me, for example, the tree, and say that water is the principle, then I will not understand why water becomes a tree. Form, therefore, plays a crucial role in the account of the tree, a role that will provide an account of the tree in the way that the material, underlying subject cannot, at least not in and of itself. Matter, it seems, cannot provide its own account, cannot explain itself. Therefore, matter cannot be a principle by itself. In addition to matter, another principle must be given, if we are to give an account of the *why*.

FROM THE GIVEN TO UNDERSTANDING: AQUINAS ON ANCIENT PHILOSOPHY

The path I have been tracing here, which leads Aristotle from an appreciation of Thales as a philosopher, because he seeks an understanding of principles, to a critique of matter as a principle, was seen already by Aquinas. For him, this path can be traversed in two modes: the one mode leading from matter as principle to form as principle, the other mode leading from matter as unintelligible to form as the intelligibility of that which is. In fact, Aquinas's entire brief history of ancient Greek philosophy is the history of bringing these two modes together.

> The ancients followed the order of human knowledge [*cognitio*] in the consideration of the nature of things. From whence, since human knowledge [*cognitio*], beginning from sense, comes upon understanding, the first philosophers [*priores philosophi*] were occupied with sensibles, and from this they gradually [*paulatim*] arrived at intelligibles. And because accidental forms are, according to themselves, sensible, not, however, substantial forms, therefore the first philosophers [*primi philosophi*], said that all forms were accidents, and only matter is substance. And because it suffices for the cause of accidents, which are caused from the principles of substance, thus it is that the first philosophers posited no other cause outside of matter; but from matter they said everything is caused that seems to appear [*provenire*] in sensible things; thus they strove to posit that there is no cause of matter and to negate efficient cause totally.[14]

Aquinas's reading of pre-Socratic philosophy follows Aristotle in several important respects. First, Aquinas identifies a kind of order in knowledge that begins with sensation and ultimately ends with understanding, that is, knowledge moves from sensible things to intelligible things. Second, these are not two orders. The intelligibles to which philosophy ultimately comes are just that intelligible element that the first philosophers seemed not to recognize within what is given to sensation. The first philosophers did not know that the substantial form is also a cause. Yet this must mean that the substantial form does not present itself to sensation and, according to Aquinas's story, must not be material. Here Aquinas pinpoints the difficulty with matter as a cause: matter is available only to sensation and thus will be the cause only of what appears in sensible things.

The failure of what Aquinas frequently refers to as an "infirmity of intellect" of the pre-Socratics has an impact on their understanding of knowledge in general. Indeed, Aquinas accuses them of holding the position that nothing is knowable. Not only is matter only available to sensation, but every body is also mobile. If matter is the only thing, then "no certitude of the truth of things could be had by us."[15] The difficulty with matter, Aquinas argues, is that, *precisely because* it is available only to sense, it makes knowledge impossible. Sensation itself is the result of the very movement of the material thing such that it can have an effect on the organs of sense. That movement, however, is at the same time the very reason why knowledge is impossible for "it has already passed by [*labitur*] before it is accounted for by the mind" (ST I, 84, 1c). Sensation is just the record of the passing of the material thing but yields no certain knowledge.

Plato, according to Aquinas, goes too far in the other direction. Plato posited "a genus of beings other than matter outside of bodies" (ST I, 44, 1c). Because these beings are of a different kind, they stand outside of motion. To this extent, Plato solves the pre-Socratic problem and saves knowledge. "In this way, he said that sciences and definitions, and anything that belongs to an act of the intellect are not referred to those sensible bodies, but to those separated and immaterial [ideas]" (ST I, 84, 1c). Yet, according to Aquinas, the price Plato pays is too high:

> But this appears false in two ways. First, indeed, because since those species would be immaterial and immobile, knowledge of motion and matter, which is proper to natural science, and demonstration through moving and material causes would be excluded from the sciences. Second, because it would seem laughable [*derisible*] that, while we seek knowledge of those things that are most manifest to us, we would be brought to other beings in the middle [*in medium*] that cannot be the substances of those bodies, since [the ideas] differ from [material bodies] according to their being. And therefore, having known those separate substances, we would not on account of that be able to judge concern these sensible things. (ST I, 84, 1c)

Here Aquinas points to the dilemma of matter. When we come to think what is, particularly that which, he insists, is most obvious, most manifest, we must think that matter is a cause and therefore we must say that it is in some way knowable. Plato moves too far away from matter and makes the philosophy of nature impossible. What is equally troubling for the philosopher who proved God's existence by means of the motion discovered among material things in the world, Plato's Ideas make a demonstration based on such moving causes impossible.

Aquinas, however, raises the stakes even further. In his demand that things yield to our knowledge, Plato *inserted* separated beings between our understanding and the things understood. In fact, Aquinas goes so far as to make explicit the errant path of Aristotle's own thought in *Metaphysics*. We set out to understand those very material things that present themselves to sensation and, therefore, are most obvious. Whatever the ultimate account of the being and being known of those material singulars turns out to be, it should make good on this initial goal. Plato, however, fails to achieve what he set out to achieve because his Ideas turn out to be the object of knowledge, that is, the things known, and, having known the Ideas, we are unable to judge anything about the material singulars we set out to know.[16]

In this way, Aquinas arrives right at the heart of the problem of materiality: it is a principle and cause and, therefore, should be part of any account of a given singular and yet it resists knowledge and stands fundamentally outside of understanding. He does, however, see a way out. In order to know, there must be an identity between what is known and the knowledge of it. However, this identity comes at the same time as a difference. What a thing is, its form or essence, constitutes its being and is, at the same time, what is known. The essence, therefore, has a double existence: it is in the thing as its being and substance and it is in the soul of the knower. Even though it is the identical essence—otherwise we could not say that we know the thing—it has different ways or modes of being in the thing and in the soul. This difference, however, is not unique to knowledge. "[E]ven in sensible things themselves we see that the form is in a different way in one sensible than in another" (ST I, 84, 1c). The example he gives, however, is not of a substantial form but an accidental one. This is, in fact, necessary, because the mode he pinpoints is precisely one of intensity. We can say that two things are white and by this we mean that the account of what it is to be white will be identical for both. However, one white can be white "more intensely" and the other less so. That his example is of an accident and, more to the point, an accident that belongs only to a material thing, might prove difficult for the conclusion he then goes on to draw. "And in this way also the form of the sensible is in one way in the thing that is outside the soul and in another way in sense, which regards [*suspicere*] the forms of sensibles without matter—as the color of gold without gold—and similarly the intellect receives the species of bodies, which are material and mobile, immaterially and without motion according to its own mode; for what is received is in the recipient according to the recipient's mode" (ST I, 84, 1c).

There is here recognition of materiality at the same time as there is a move away from it. The analogy turns on a feature that belongs to matter, namely intensity.[17] Aquinas is searching for a way, literally a *mode*, so that he, and Aristotle as well, are not committed to Platonism. The most obvious modal difference is intensity in relation to accidents. That sweetness or color, for example, are susceptible to degrees is not only manifest, obvious, but the very clarity of the degrees of qualities belongs to matter alone. It is not manifest that being or human is similarly susceptible of modes of intensity. To say that one thing is sweeter than another is to name a twofold material difference: a difference in the material constitution of the thing and a difference in the material effects the thing has on the organs of sense. Modality understood as intensity, therefore, seems to belong most of all to material things.[18]

Aquinas brings this feature of material accidents into analogy with that which is other than material. The "ways" that belong to more sweet and less sweet, more white and less white, are just like the way in which the form of the sensible in that material thing differs from the way in which it is in sense "without matter." Aquinas assumes that there is a difference in modes, in ways, of which intensity is just one kind, one is tempted to say, "one mode." Another mode of modality is the difference in the way in which the form is "in" sense and "in" the thing sensed. Since we can obviously say that two different things are both gold, sense must receive the goldness independent of the matter without which there would be no gold. Already here there is a step away from matter such that sensation itself is moving toward the immaterial. To the extent that sense is of gold and not of this thing that is gold, sense already moves from the "manifest" thing given to it toward something intelligible. This is marked by Aquinas's use of the term "form" to refer to gold without a gold thing.

This first step away from materiality makes possible the second: the intellect can receive the species of a body according to its own mode: immaterially and without motion. Even though any body, by definition, is material and moved/moving, its species, which both Aquinas and Aristotle understand to be the form/essence of the thing, can be identical, can be the same, even when received without matter and motion. According to Aquinas, there is a "what" that is received by the soul. The *what* never alters, moves, or changes. What changes is *merely* the *way*, the mode, in which it is received. The form/essence is the same whether *in* the thing as a principle of being or *in* the soul as that which is known. The only difference is *merely* one of the *way*, the *how*, in which it is. However, if the *how* is irrelevant, then how could even this position save the initial hope that we could know the material singular? Does this shift toward a merely modal difference not signal that the soul has turned its back on matter?

FROM PRINCIPLE TO ESSENCE IN ARISTOTLE

Since Aristotle argues in both *Metaphysics* and *Physics* that matter is a cause, it certainly plays a role in explanation, that is, in giving an account, a reason, of that which does not present its own reason. Yet there is an oddness to matter that is marked by Aristotle:

> Since the causes are four, it is the task of the physicist to understand all of them; and as a physicist he should state the *why* by referring it to all of them—the matter, the form, the mover, and the final cause. The last three often amount to one; for both the whatness and the final cause are one, and the first source of motion is the same in kind as these (for man begets man), and, in general, this is so in the case of a movable mover. (Phys. 198A22-28)

Here Aristotle marks the oddness of matter by showing that it, in a sense, is a principle and a cause that is different from the other three. When a human begets a human, it is begotten by a human in order to be a human. There is, then, an identity in the other three causes such that matter stands outside the circle cast by these causes. It turns out that there are only two principles: whatever is identical in agent, form, and telos, and matter. The four causes turn out to be only two. The coincidence of form/essence in the agent and telos leaves matter out of the account of the *why*, which now turns out to be the same as the what-it-was-being, the essence. Matter can only be a principle, therefore, of the *that* and does not bring forth the *why*. Moreover, matter here begins to take on the role of the merely passive. From passive, it is a small step to potency and potency itself is a small step from nonbeing. If the agent introduces a form so that the result is essentially identical to the agent, then this all is *done to* the matter. Matter begins to recede the moment it is understood as merely undergoing action and playing no role in the being of the material thing.

In Book Zeta of the *Metaphysics*, Aristotle turns his attention once again to matter, this time in relation to the question of what is that which is. That οὐσία is the most obvious answer to the question "What is that which is?" is itself quite obvious. In fact, more literally rendering Aristotle's text, we could read him as arguing, "Given that what is is spoken of in as many ways as this, it is obvious that the first thing that is is the what it is, the thing that signifies the is" (Meta. 1028A14). We could recast this to point out the obviousness: "If you want to understand what is, the first thing that is is the what, and that signifies 'is.'" The term that prevents this from being a maddening repetition of forms of the word "to be" is "the what," τὸ τί. Aristotle shows the importance of this as he turns his attention now to the question, "What is οὐσία?"

He begins this account by naming four candidates: "(a) that which it was to be that thing [τὸ τί ἦν εἶναι], (b) the universal, and (c) the genus, and there is also a fourth item in this list, namely (d) the subject" (Meta. 1028b34–36). We can note, first, that Aristotle again marks an oddity that belongs to matter. The underlying subject (ὑποκείμενον) is something that is added as a supplement to those items that seem more readily to count as substance. Aristotle takes up this question of matter first. Substance seems to be first of all either the matter, or the form, or the composite. Aristotle indicates that by "matter" he means "the bronze of the statue." It is, however, the second member of the composite that proves interesting. Aristotle begins by using the term μορφή, that is, shape, and then indicates that, by this, he means the ἰδέα, the idea of the composite. He then goes on to say that this is prior to and "more a being than" the matter, and therefore it will be prior to the composite. In this claim, Aristotle substitutes a third term as equivalent to both μορφή and ἰδέα: εἶδος.[19] It is not clear, at least immediately, what it could mean for one being to be "more a being" than another. This notion seems related to Aristotle's insistence that the *what* is the first thing that is. The matter, as a result, must have a secondary kind of being when compared to the form.

This account turns out to be insufficient, however, because "it is in itself obscure, and secondly, on this account as it stands, *it is matter that turns out to be substance*" (Meta. 1029a10). That is, the account must be revised, Aristotle seems to worry, lest matter be considered a thing that is. He goes on to insist that it cannot be that matter is a substance, "because surely what we really think to be the most central features of substance, surely, are separability (χωριστός) and thisness (τόδε τι)" (Meta. 1029A28). Matter, for Aristotle, cannot be a substance because substance is both separable and "a this." The composite of matter and form may seem to be a substance because it is a this (τόδε τι) and may even be separable. However, as we have already seen, the composite has these characteristics not originarily but as a derivative of the fact that form alone already has them. So on the one hand, matter is too apparent (φανερός), too straightforward, too immediate to count as substance, while the composite, on the other, is too late, too derivative. It leaves, then, only the third, the *εἶδος*/ μορφή as a candidate for substance. Yet the being-too-apparent of matter disqualifies it for being only on the condition that being must, in order to "be more a being," be intelligible.

The immediacy of matter turns, for Aristotle, precisely on the question of separability. As he indicates later, the characteristic of separability applies not to what we might call real separability, in the sense that it can be really separated from another, but *logical* separability (κατὰ τὸν λόγον), that is, its account is a separate one (Meta. 1042a29f). In addition, matter only counts as a *this* in a qualified sense. Since, Aristotle argues, perceptible substances are unquestionably substances, we have to turn once again to the question of matter, since what characterizes all perceptible substances is that they have matter (Meta. 1042a26). Here, matter is defined as "what is not a possessor of thisness in actuality but is a possessor of thisness in potentiality" (Meta. 1042a26f.). Since matter is only potentially a *this*, it must rely on something other for it to be a *this* actually. If being a *this* is a necessary characteristic of substance, then matter cannot count, in an unqualified way, as a substance, since it needs something other than matter in order to have that characteristic. The designation of matter as merely potentially a *this* pushes it further from what is, from the actual. Again, the inability of matter to be thought forces it out of what counts as real.

There is, however, another way of looking at the material substrate, viz., as the account (λόγος) and shape/form (μορφή). By taking the matter not as matter but as the form of that substrate—taking it as wood rather than the matter of the bed—matter ceases to do its crucial work as matter. In this way, we turn away from the materiality of matter by bringing its form to the fore. Only in this way, Aristotle argues, can we say the substrate has thisness and separability in *logos*. There is, therefore, a difference between matter as matter and matter as already something. If we take, for example, a house, we can say that it has matter in that it is made of bricks, beams, et cetera. Here, however, the matter is itself a substrate each one of which is also something that has matter and form of its own. That is, we can understand that the bricks are some *this* unto themselves and independently of the house. Viewed in this way, the substrate is something that has thisness and is separable in its account. We can say what a brick is without ever saying anything about a house, and thus the account of it is separate from the account of the house. However, a problem immediately arises. The brick has matter of its own that, perhaps, also has shape/form, and therefore is itself a this and separable. In other words, to view the material substrate as itself a substance does not ever bring to the fore what we might call the materiality of matter, that is, matter as such without regard to form/shape.

THE PROBLEM OF FIRST MATTER

There is here an analytic difficulty. On one hand, Aristotle points to matter and substrate (ὑποκείμενον) as a principle that belongs to the *why* of what is. On the other hand, whenever we turn our thought to matter, what emerges is not matter in its materiality, but a form. The matter of the statue is bronze, that is, something that we grasp in its what-it-was-being. As soon as the form emerges under our thoughtful gaze, the bronze as matter slips away. However, even while we grasp the form "bronze," we also recognize that it has its own matter. As we turn our attention to the copper and tin, we repeat the same problem again. There is always a function of matter, which I have been calling materiality, that is other than the form and is, consequently, not grasped when one grasps the form of the matter of a given particular thing. The very indication that there is matter begins this analytic process that, if it goes without end, will not only make matter unknowable but will also call into question the very notion of form.

From his medieval interpreters on, it was thought that Aristotle solved this problem with the notion of first or prime matter. If everything comes to be from some preexisting matter, then that must be the case also for the primary forms of matter, namely the elements. That is, if all matter is resolvable, analytically, to the basic forms of earth, air, fire, and water, then it seems natural to ask about the matter that constitutes these basic forms of matter. While this might solve the problem of infinite regress, it faces two major difficulties of its own. First, Aristotle does not seem to discuss this prime matter whenever he comes to face just this difficulty. Second, any notion of first matter would be the notion of matter without any form whatsoever. For Aristotle, that which is without form would be that which does not have a what-it-was-being and, therefore, is not actual in any way. Prime matter, then, turns out to be a performative contradiction. If it exists, it does not exist.

For Aquinas, *prima materia* is a kind of analytic residue in that it is what is left over after analyzing the basic, elemental bodies (earth, air, water, fire). As the "ultimate resolution" of natural bodies, it lacks all form. Since, according to Aristotle and Aquinas, knowledge is through form, prime matter is knowable only by analogy: "prime matter is related to all bodies as wood is related to the bed."[20] That prime matter has no form means that there is nothing to understand, know, or think in relation to it. It is "grasped" only by analogy to generation, destruction, and alteration of substances whose matter is itself "formed," that is, has its own essence. There is, because knowledge here is only by analogy, a "something there I know not what," that belongs to prime matter.

What Aquinas suggests here regarding prime matter, however, must be extended to matter as such. Aquinas refers to prime matter as "that in which ultimately stands the resolution of natural bodies" [*illud in quo ultimo stat resolutio corporum naturalium*]. It appears to be "ultimate" because if we want to analyze the matter of a bed—and we should note that the analogy is not to a *natural* body—we simply turn to wood. Yet, as I have argued, wood is itself already a reference to the essence and form of the matter. Because this analysis turns to the essence and form, it, in fact, misses the matter that is the matter of the bed *as matter*. If matter is distinct from form, if form is not matter and matter is not form, then there is the matter of the bed that is in operation other than as having the form wood. It is not insofar as wood has an essence and form that it is matter, otherwise any pile of wood would be a bed. So the very characteristics that Aquinas ascribes here to prime matter—its being without form and, consequently unknowable in itself—must belong to every material thing. This is what I have been naming "materiality," that is, the mode of being of matter that is outside of and indifferent to form. Prime matter is not, therefore, an incidental feature of the matter that is considered to belong to elemental bodies, it is the "essential" feature of matter as such; it is materiality.

The analogical grasp of matter, of materiality is not, for Aquinas, a result of the finite, embodied, human intellect. Not even God can know prime matter directly. This is a curious moment, for Aquinas frequently argues that God is to the universe as an artisan is to artifact. As I indicated in Aristotle's distinctions that bring wisdom to light, the artisan, the one with skill, should have skill that ranges over the materials that will bring about the final product. It would be remarkable if the sculptor had no idea of the bronze or the stone. It would be even more remarkable if the plumber had no knowledge of water. But God is the cause of all things, and thus is also the cause of prime matter. So God must know, "in some way" [*aliquo modo*], prime matter. Yet Aquinas goes on to argue that "if we speak properly of an idea, it cannot be posited that prime matter has, per se, an idea in God distinct from the idea of form or of the composite; because an idea, properly called, regards a thing according to what is producible in being; matter, however, cannot go out into being without form, nor the converse."[21]

Here Aquinas comes dangerously close to insisting that God also has analogical knowledge of prime matter. Since there can only be an idea, which Aquinas has defined as God's knowledge of what is to be produced, of the form or the composite, and since prime matter, by definition, has no form and is not a composite, God could have no knowledge of it. What God knows, as Aquinas goes on to show, is that there is something from which the first composite is made. Again there is a "something-I-know-not-what" that belongs to prime matter. However, once again Aquinas cannot restrict this just to prime matter. Even for God, there must be something about matter as other than form (and therefore other than thought) that comes to the fore in prime matter but is not exclusive to it.

MATTER AS ERRANCY

It turns out, therefore, that matter is an odd sort of principle. On the one hand, perceptible substances seem most of all, or at least most obviously, to be substances and what they all have in common is matter. On the other hand, matter as matter cannot account for the reason why a substance is, is what it is and, therefore, matter as matter cannot pertain to what is most substantial for substances. Aristotle seems to give two arguments that move in just these two opposing directions. For those things that are constituted, at least in some part, by their materiality, matter must be a principle, since otherwise a full account of them would be impossible. However, matter as matter seems to be that of which there is no account, for every account will be somehow related to the form/shape and, therefore, push further away matter as a principle. It is unique among all the principles that Aristotle discusses in that in doing its work of bringing about and organizing, of commencing and commanding—that is, in being an *arche*—it resists and confounds knowledge, understanding, and wisdom.

Aristotle's critique of Thales, therefore, does not amount, in the end, to a rejection of his original insight. To the extent that material substances are substances in an uncontroversial sense, materiality is an irreducible principle of those substances. In order to *think* those substances, therefore, one strives to think that principle that is their materiality. However, the materiality of matter recedes from that very thinking. There is an element that will always stand as something more, something supplemental to the thinking of that material substance. On his way to trying to give an account of what is, Aristotle seems to have fallen back into the very position for which he critiques Thales, namely, there is a principle that is both necessary for thinking what is that, at the same time, recedes from thought. Indeed, Aristotle seems to implicitly recognize a central feature of matter as matter: it is, simply, other than thought. Aristotle is forced to this position, it seems, by his acknowledgment that the conditions that allow something to be thinkable both require and refuse matter. In this way, Aristotle's path to matter brings to the fore the basic feature of matter: it is nonthought, its being is as an *other* to thought.

Yet the recognition that matter is both the aim of thought and receding from thought is not Aristotle's final word. Indeed, he almost immediately turns away from this insight back into thought. As we already saw, one of the candidates for what could count as substance was the "what-it-was-to-be-this" (τὸ τί ἦν εἶναι), that is, the essence. When it comes to essence, the primary characteristic is that it is καθ᾿ αὑτό, according to or through itself. If we look after what-it-was-for-you-to-be, being musical would not belong, since you are not musical through or according to yourself (Meta. 1029b10). The problem with accepting something like "musical human" as the what-it-was-to-be-this is, once again, related to being a this as a feature belonging necessarily to substances: "a what-it-was-to-be-that-thing is the same sort of thing as a thing with *thisness*. And a thing with thisness is just what we do not get whenever one thing is said-of another" (Meta. 1029b–1030a).

In investigating whether essence (the what-it-was-to-be-that-thing, τὸ τί ἦν εἶναι) is a substance, Aristotle has now uncovered a distinction that seems to have already been operative in his notion of a *this,* namely that whatever is a this cannot be something that is said of another. When we take such a thing, for example a white log, we do not take its substance, because to be a white log is not just to be white. White is said of a thing that has thisness but in itself it is not something with thisness—or only in a derivative and, therefore, deficient way. Predication indicates something crucial in that it shows that what is predicated of another is not itself something with thisness and therefore is not something that is an essence.[22] This shows, according to Aristotle, that there is a difference in giving an account of something that is a this and something that is not:

for anything that is a this, the account will be a definition. If we ask what it is to be a human, the account given will be of something "first" (πρῶτος, πρότερος). Since Aristotle contrasts this with the case of something like "white," we can assume that what is first will not be something that can be predicated of another. If we were to give a definition of "white," something else will occur in that definition because there is always something (i.e., a surface) that is white. For this reason, white cannot count as an essence because it will always be said of something that will be a this and therefore is not itself a this.

Up to this point, Aristotle has been focusing on the notion of what could be a τόδε τι. While certainly τόδε brings an indexical, that is, a finger point, to language, we have not realized the importance of the τι, that is, "something." The pointing brings to the fore that we are trying to bring to language a thing standing there in front of us. Yet, what Aristotle also recognizes is that the thing standing there in front of me is nothing if it is not some kind of thing. Its being at all is wrapped up in what it is. This is not to confuse the "something" with what Aristotle frequently refers to as τοιόνδε, a "such-a-kind." He seems to distinguish between a kind of generality or universality that belongs to something like a genus, for example, animal, from the kind of generality that belongs to the species below which there are only individuals.[23] This acknowledgment that every individual is something, that when confronted with this individual we can say what it is, puts Aristotle in a difficult position. As Yi points out, Aristotle does seem to contrast sharply the being-a-this that belongs to substance with being such-a-kind: "These, then, are the puzzles with which we must deal in connection with the principles, and also whether they are general or, as we claim, particular. For if they are general, they will not be substances (for none of the common things signifies a τόδε τι but a *τοιόνδε*, whereas substance is a τόδε τι)" (Meta. 1003a5–14).[24] Yet if substances have no relation to generality, they will not be knowable (Meta. 1003a16).

The difficulty, therefore, is that if substances have no regard nor relation to something general, they will not be knowable, yet, if they are themselves universal, they will not be substances because they will not each be a τόδε τι. This, however, is nothing other than the very materiality of matter, that is, what makes matter as such, namely that it is other than thought. As a principle, matter always recedes from knowledge, understanding, and thinking because thought in all these ways only ever encounters the formal, essential, and therefore general, aspect of the material thing. Yet Aristotle never doubts that matter is a crucial principle of, at least some, things that are. There is here an obvious aporia in that, as a principle, matter must be understood if we are to understand any material thing, and yet matter as matter is never that which is understood. While Aristotle, in the trajectory that has just been traced, will ultimately think the being and, therefore, the reality of the concrete, material, singular in terms of its essence, that is, in terms of what is thinkable in it, it is fruitful to pause here to note what comes to the fore on the way to that ultimate subsumption. In his thinking through why matter cannot count as substance and why essence ultimately is the only thing that can be a substance, Aristotle implicitly recognizes a difference between what is knowable, that is, the thing that can be an object[25] of knowledge, and that knowledge itself.

Aristotle's critique of Thales, in fact, played out along the difficulty of thinking that which presents itself immediately. The critique was not a rejection of the twofold direction of Thales's thought. It is a critique neither of Thales's insight that what presents itself is not as it appears and therefore its manifestation is not the whole story of its reality, nor of his insight that what thought achieves in relation to what appears is based on conditions that belong to thought and not to what appears as it appears. This is the very thin line of matter, of the materiality of matter. According to Aristotle, Thales recognized two main points and this twofoldness seems to be the necessary beginning of philosophy: thought is not *causa sui* in that it is always *about* something that makes a claim on thought and, as a necessary consequence, is other than thought; and what presents itself is not immediately what it appears to be and, therefore, thought is required in order to grasp the reality of what appears.

This twofold recognition is what constitutes materialism. That is, materialism is necessarily caught between what is given (and that will always be material) and its mode of being given (and that is never graspable except by thought). Aristotle's critique of Thales does not reject Thales's fundamental insight that there is a need, indeed a materialist need, to come to understand that which presents itself materially. That is, Aristotle insists, along with Thales, on the "principle distinguishing between the *real* and *thought*."[26] However, Aristotle's critique of Thales hinges on his fundamental principle that there is no ἐπιστήμη, no knowledge, of a singular. This is not to equate materiality and singularity, though we shall see that there is a crucial connection. Rather, just as matter in its materiality is not graspable by thought, so too the singular is not graspable by ἐπιστήμη. That is, Aristotle's overriding concern is not to expose the irreducibility of matter in relation to what is but the thinkability of what is. This, in the end, is the essence of his critique of the early natural philosophers. Matter is a principle of what is, yet it cannot be the end of the account because it cannot, as such, be known or thought. In this way, matter occupies the same systematic position as the singular.

Aristotle argued that matter cannot count as a substance because it failed to be either separable or a this, let alone both. This led him to the conclusion that form is more substance than matter and even the composite of form and matter. But his ultimate conclusion is that it is the essence—the-what-it-was-to-be-this—that is substance. That is, "thisness," τόδε τι, is not that which lies at the end of a finger, an individual that can be pointed to but, rather, to the species of a genus (Meta. 1030a).[27] For Aristotle, the essence as a species of a genus is consistently and fundamentally linked to the question of a definition. Only that can count as substance that is capable of definition and only a species is such a thing, since, it seems "a definition is an account of the what-it-was-to-be-that-thing and a what-it-was-to-be-that-thing either belongs only to substances or belongs to them in the fullest sense, primarily and without qualification" (Meta. 1031a10–15).

This is the moment at which, traditionally, the question of the realism of universals is raised, along with its related question concerning the "principle of individuation."[28] Yet the issue is, in a sense, already decided. To the extent that knowledge is the priority, to that extent some level of generality must be involved, as Aristotle himself acknowledges: "that the particular thing itself [ἕκαστος] and the what-it-was-to-be-that-thing are non-accidentally one and the same . . . can also be shown by the argument that knowledge [ἐπιστήμη] of some particular thing is constituted by knowledge of the what-it-was-to-be-that-thing" (Meta. 1031b19–20). This conclusion, however, was already implicit in Aristotle's critique of Thales in that Thales argued that the account of what is (water) is that of which there is no knowledge and thus an account is not given. Yet is this not the other side of the coin whose face is matter as other than thought? If matter cannot be thought, and if the thing is thinkable, then thought will reach something other than matter.

Aristotle immediately takes up this question of matter in relation to the essence. He recognizes that when anything comes to be, either by nature or by τέχνη, there are three, as it were, moments: (1) that which is produced, (2) that by which it is produced, and (3) that from which it is produced (Meta. 1032a13ff.). The "from which" is, according to Aristotle, the matter. Yet the matter disappears in a kind of silence: "And in these cases that by which the production occurs is one of the things that exist by nature, while that which is produced is a [human], a plant, or something else like that" (Meta. 1032a18f.). Without comment, Aristotle moves away from matter, from the "from which," to the conjunction of what is produced and that by which it is produced—the form of the producer and the form of the produced. Disregarding the "from which," that is, the matter, in any coming to be there is an identity: that by which something is produced is the same as that which is produced: "This nature is of a single specific form, being the same in a different instance, in the way in which it is a [human] that begets a [human]" (Meta. 1032a24–25).

In this account of generation, matter plays only a passive and, in fact, transitive role. While Aristotle here recognizes that generation cannot be understood without the "from which" it takes place, the matter cannot play a role in the philosophical account of generation. Generation turns out to bring forward an identity between that which is produced and that which produces. The identity is a "formal" one and, precisely because it is formal, both requires matter and subsumes it under this identity.[29] That form that belongs to both the product and the producer is identified with the essence, with the what-it-was-to-be-that-thing: "Now, things are produced from skill [τέχνη] if the form of them is in the [soul] (and by form I mean the what-it-was-to-be-that-thing for each thing and the primary substance" (Meta. 1032a–1032b).

The introduction here of the requirement that in τέχνη the form [εἶδος] must be in the soul of the producer indicates something crucial about the formal identity of producer and produced. The form/essence is thinkable, that is, it pertains both to thought and to the being of the thing. In cases of natural generation, the form/essence is the very form/essence of the producer and generation is just the production of that form in something else. Natural generation just is reproduction or the assimilation of some matter to the form of the producer. Artificial generation also has this same structure, with the exception that the form is not the form *of* the producer but a form *in the soul* of the producer. Τέχνη, therefore, shows that the form/essence is not itself material, since it can be in the soul. "The upshot is that in a way health is produced from health and a building from a building, the building, namely, that has matter from that which does not. For medicine and architecture are respectively the forms of health and a building, and I say that this form without matter is the what-it-was-to-be-that-thing" (Meta. 1032b13f.). The conclusion Aristotle reaches here, however, cannot be limited to τέχνη. Rather, it shows that the essence, the what-it-was-to-be-that-thing, is without matter and, therefore, thinkable, separable in the way that it can come to be in the soul without matter.

Something has gone awry in this path I have been tracing. The critique of Thales started out with the very notion that we are trying to *think* what is, to give an account of it. As Aristotle points out, the thing that is in the most obvious sense is that which is available to perception and, therefore, a material thing. Matter was seen as an irreducible element of that thing being at all. Yet it cannot count as substance, it cannot be the very being of the thing, because it is not separable. The separability criterion has turned out to be separable in thought, separable in being in the soul without the matter. Essence as the very being of what is means that matter must fall out of the account, even while it is necessary and a principle of what is. Here Aristotle indicates that what is thinkable, what can be in the soul, is only the form without matter. Working in the other direction, this must mean that matter as such does not pertain to thought—it cannot be in the soul at all.

The main alterity between matter and what is thinkable continues even further, with the same oddity that while matter is an ineliminable element of what is, it is not part of the account of what is because the account is that which is given by thought. We certainly can ask for an account of matter, we can ask, "What is it?" Yet in each case the account, the answer, will be a form: bronze, brick, bones, and so forth. When we ask after what is, the answer is a form: a statue, a house, a human. The matter becomes, in this case, an additional qualification. The statue is because it is a statue, yet it cannot be a statue without some matter. The thing that is, however, is called a bronzen statue or a brick house. In each case, the matter now takes on the character of the superfluous, indeed of the accidental—a house is a house whether it is made of wood, bricks, or even straw. So even though "a bronze circle has its matter in its account" (Meta. 1033a10), the matter is not the what-it-was-to-be-that-thing.

Surely Aristotle has gone off the path. As I noted above, the task he set was to think what is. His critique of Thales fell on this side of thinking. That is, his critique exposed how Thales, a philosopher, failed to offer a principle that is adequate to thinking. Aristotle needed, then, to set him straight, to put him back on the path, the path of thinking. Yet as Aristotle himself reconstructed that path, it turned out that it did not lead to what is but only to what is thinkable. What is most of all is what is called substance. Substance, in its turn, is essence, that what-it-was-to-be-that-thing. Essence stands as both the very being of what is and what is thinkable. However, as thinkable, essence is other than matter. Matter, the very materiality of matter, is other than what is thinkable and, following Aristotle's path, is therefore not essential to the material thing. And thus Aristotle's path has missed its destination: that which is.

The errancy of Aristotle's walk, however, indicates something crucial in the attempt to think matter, to think that which is material: thought and matter stand in a fundamental alterity to one another. This fundamental alterity presents a difficulty for any attempt to think that which is irreducibly material because materiality escapes thought. In other words, the difficulty is that thought pushes away its "object."[30] This difficulty is relatively easily surmounted if one posits, as I have been arguing Aristotle does, a fundamental identity between what is thinkable and what is. However, I have also been showing how, for Aristotle, matter in its materiality is at the same time irreducible. It is a "something-I-know-not-what," a *tertium quid*. In this way, Aristotle presents the challenge of thinking of matter. In the very thinking, matter is continuously pushed out of reach, and yet, if thinking is to be *of* something, it must be a thought of that which is material. Aristotle's errant path, therefore, is nothing less than the errancy that always belongs to thinking of matter.

NOTES

1. I am indebted to Sean Kirkland for following this argument more carefully.

2. Aristotle, *Metaphysics*, trans. Hugh Lawson-Tancred (London: Penguin Books, 1998), 6. 928b1. Hereafter, this text will be cited as Meta. followed by the Greek page and line numbers. Unless otherwise noted, all translations of the *Metaphysics* are those of Lawson-Tancred.

3. Here, again, it should be noted that we have no "text" of Thales and, therefore, what Aristotle insists is Thales's position is, at the same time, the construction of Thales's philosophy.

4. As, perhaps, is already clear, Aristotle refers to and distinguishes between a variety of capacities that, in general, fall under the English term "knowledge." I am not entirely happy with the various connotations that the term "cognitive"

has in contemporary philosophy. For this reason, I prefer the more general and perhaps neutral, if infelicitous, term "gnoseological."

5. The relation between sensation and particularity/singularity will become more central as I follow Aristotle's path here. It will come back again when we turn to Scotus (chapter 3) and Hobbes (chapter 4).

6. I do not engage here in a discussion of the relation between ἀρχή and λόγος. Whatever the case with that relation, it is clear that Aristotle begins the *Metaphysics* with an understanding that whenever there is an account (λόγος), that account will never come to rest until it reaches an ἀρχή. This is not simply a question of putting that which is into language but, as Aristotle has already been arguing, the understanding of what is in terms of *why* requires an account and that will ultimately end in an ἀρχή.

7. I use "because" as a formal or empty answer to the question "Why?" What Aristotle also raises here is the question of how, in what way. In chapter 5, I return to this as a question of mediation.

8. "Nor is practical wisdom concerned with universals only—it must also recognize the particulars; for it is practical and practice is concerned with particulars." Aristotle, *Nichomachean Ethics*, trans. David Ross (Oxford: Oxford University Press, 1984), 1141b.

9. Since matter is one of the four causes, it surely does play an explanatory role for Aristotle. However, its role is diminished by another principle, ultimately form, that provides a fuller account of the *why*. As I discuss in chapter 5, this leads to the paradox that matter must be a principle but it cannot be a principle.

10. One of the central insights of the *Metaphysics* is rather blunt and obvious: there is nothing that is not something. What Aristotle draws from this is the notion that *what* something is turns out to be its very being. This insight, however, leaves out the role that matter plays in a thing's being at all. I will trace the withdrawal of matter from Aristotle's account of what it means for anything to be in what follows.

11. As I argue in the next chapter, the problem of immediacy is the central problem Marx, Adorno, and Althusser recognize in relation to materiality. That is, the immediacy of what is given does not present the *how* of its givenness. While Aristotle's interest in understanding and wisdom pushes him to the *why*, the main problem in relation to matter might be the *how*. I return to this in chapters 2 and 5.

12. Aristotle, *The Politics of Aristotle* (Clarendon Press, 1885), 21, 1259a8ff.

13. Hippocrates Aristotle, *Aristotle's Physics* (Grinnell, IA: Peripatetic Press, 1980), 21, 190b18–21. Hereinafter, this text will be cited as Phys. followed by Greek page and line numbers. Unless otherwise indicated, all translations are those of Apostle. On the question of the search for principles in *Physics*, see Johannes Fritsche, *Methode und Beweisziel im ersten Buch der "Physikvorlesung" des Aristoteles* (Frankfurt: Verlag Anton Hain, 1986).

14. Thomas Aquinas, *De Potentia Dei*, Question 3, article 5. The text can be found at http://www.corpusthomisticum.org/qdp3.html.

15. Thomas Aquinas, *Summa Theologiae*, Pars I, quaestio 84, articulus 1. This text will be referred to as ST followed by part, question, and article numbers.

16. As I discuss in chapter 3, this is what leads Adorno to the claim that there is no metaphysics in Plato because he fails to show the mediation between this world of material singulars and the being of Ideas.

17. Aquinas turns to what I am calling here "intensity" because, it seems, intensity is one way to grasp a scale of "more" and "less" *of the same thing*. That is, intensity marks a difference within identity.

18. Duns Scotus, however, argues that there are "intensities of being," primarily "finite" and "infinite." This is related to his argument that "being" is univocally predicated of God and all other beings. The difference between God and this rock, as far as the notion of being is concerned, is only a matter of mode, particularly intensity. On this question, see Francis Catania, "John Duns Scotus on Ens Infinitum," *American Catholic Philosophical Quarterly* 67 (1993): 37–54.

19. This is, to be sure, a difficult passage to render into English, particularly as the medieval, Latin tradition often rendered Aristotle's *morphe, idea,* and *eidos* simply by *forma.* This passage gives a good reason for this option. However, it has a tendency to conceal *both* the differences among these terms *and* their interrelations in Aristotle's texts. I have here stayed away from Lawson-Tancred's translation, although it attempts to keep these terms separate.

20. Thomas Aquinas, *Scriptum Super Sententiis,* accessed October 21, 2014, http://www.corpusthomisticum.org/snp2012.html#4580 Liber 2, distinctio 12, quaestio 1, articulus 4, corpus. Hereafter, this text will be cited as Super Sent., followed by book, distinction, question, and article thus: Super Sent. II, d. 12, q.1, a.4c.

21. Thomas Aquinas, *Quaestiones Disputatae de Veritate,* accessed October 21, 2014, http://www.corpusthomisticum.org/qdv02.html#52314 Quaestio III, articulus 5, corpus.

22. In *Metaphysics,* Book 9, chapter 10, Aristotle returns to his claim that "the true" is one of the ways in which being is said. There, he seems to draw a distinction between the being-true of "white log" and, for example, "human." The prior has to do with predication while the latter has to do with essence. On this, see Richard A. Lee and Christopher P. Long, "Nous and Logos in Aristotle," *Freiburger Zeitschrift fuïr Philosophie und Theologie.* 54, no. 3 (2007): 365ff.

23. In this way, I disagree with Jiyuan Yu, *The Structure of Being in Aristotle's Metaphysics* (Springer, 2003), 125ff. The fact that Aristotle distinguishes between *toionde* as a level of generality and *tode ti* does not mean that "this something" is not constituted in its being by some generality, that is, by essence. I agree that anything that is a "this" (*tode ti*) is singular. However, its being comes to be identified with its essence, its what-it-was-to-be-that-thing, which is not singular but belongs, at the same time, to any other individual having that essence.

24. Ibid., 126.

25. I do not use the term "object" in its Kantian and post-Kantian sense. Aristotle, too, recognizes the need for thinking that emerges only when a capacity of the soul (sensation or thinking) meets the world in the form of a proper *antekeimenon,* that is, object. On this issue, see Lawrence Dewan, "Obiectum: Notes on the Invention of a Word," *Archives d'Histoire Doctrinale et Littéraire de Moyen Age* 56 (1981): 37–96.

26. Louis Althusser, *Reading Capital* (New Yok: Verso, 2006), 86–87.

27. While I cannot argue the point fully, it seems to me that Aristotle's position is not fundamentally altered by his introduction of actuality, ἐνέϱγεια. In fact, I would argue that this only exacerbates the basic problem as now essence is read as the actuality of the thing leaving matter as *mere* potentiality.

28. We return to this in chapter 3 in relation to Duns Scotus.

29. Johannes Fritsche, "Genus and To Ti En Einai (Essence) in Aristotle and Socrates," *Graduate Faculty Philosophy Journal* 19/20, no. 2/1 (n.d.): 163–202.

30. Yet again, "object" must not be understood here as something other than the thinking subject, but rather as that which thought is about, that toward which it strives.

TWO

Marx, Althusser, and Adorno

The errancy of the thought of matter that Aristotle exposed seems to be almost intractable. If matter as such, matter in its materiality—that is, not in regard to form or essence—is other than thought, then any attempt to think matter will, it seems, fail in the same way as Aristotle's thinking of matter failed. There is a methodological problem at the most basic level. Thought is always *of* something, yet when that something is matter, thought lacks its object, lacks that *of* which it is. As Aristotle's critique of Thales showed, matter cannot be a principle, since it cannot be thought, and yet must be a principle,[1] since it is an ineliminable element of what is, whenever that which is turns out to be material.[2]

The path that Aristotle walks, the errant path, Althusser names "empiricism."[3] For Althusser, empiricism is the name of a process that occurs between thought or the thinking subject and the object. Empiricism is the notion that both thought and object "predate" the moment or circumstance of knowledge. That is, thought as such is given before any object and the object as such is given before any thinking subject comes to engage it. While the obvious duality is that of the real object and the knowing subject, Althusser argues that this duality is derivative of a more fundamental duality that is posited, by empiricism, already *in* the real thing that is known, that is, it is a "function of the real object."[4] The issue for this model of knowledge is how the thought of the object can be the thought of the real object, that is, how knowledge can claim to be *of* what actually is.[5] The key to this identification of the thought and the real is, for Althusser, abstraction. "To know is to abstract from the real object its essence, the possession of which is then called knowledge."[6]

The process that Althusser calls "abstraction" was the very move that was traced in Aristotle's thought above. The soul comes to think the thing in its essence and not in its given materiality. As we saw, the justification for this errant move was Aristotle's insistence that what is thinkable of any given thing just is that which gives it to be, that is, its essence. Yet Althusser points out that this necessitates a duality, though not the dualism of thinking subject and its object, but a duality in the object itself. He shows, on the contrary, that if there is the thing and its essence, then in the object of knowledge itself there must be this duality. "In every case, this *separation*, in the real itself, of the essence of the real from the dross that conceals the essence, imposes a very special representation both of the real and of the knowledge of it, as the very condition of its operation."[7] This duality means that there must be, in fact, *two* essences, an essential one and an inessential one. The essence is, after all, the essence *of something*. That of which it is the essence, therefore, turns out to be essential to the essence. However, in relation to the extracted essence, what is essential to it is inessential. Just as the gold miner finds gold only after extracting it from the actual thing found, so too empiricism finds the essential essence only by extracting it from the actual thing found.

The real is, therefore, twofold and the essence is just one of the parts. The other part is eliminated "by a whole series of sortings, sievings, scrapings and rubbings."[8] While for Aristotle and other "empiricists," there is an achievement here, namely the grasping of that which is the being of the thing, Althusser brings to the fore the other side, namely, that such processes attempt to "leave no trace in the extracted part, every trace of their operation is eliminated along with the part of the real they were intended to eliminate."[9] If the essence emerges only as a result of a process, that is, if the essence does not show itself immediately to sensation, then one would expect that process to leave a mark on its result. Yet this attempt is in vain; for such is the nature of a trace. The process does leave a trace. As Derrida argued, "The trace is not a presence but is rather the simulacrum of a presence that dislocates, displaces, and refers beyond itself."[10] In this way, the empiricist process attempts to do its work without leaving behind a mark of its operation. However, the operation of abstraction is on the real object and part of it is left behind when the essential part is taken up as knowledge. As Derrida goes on to say about the trace, there must be retention of "a mark of what [is] lost or put in reserve, set aside."[11]

For Althusser, the trace of the extraction of the essential from the real object occurs "in the conditions of the operation; to be precise, in the *structure* of the *real object* from which the knowledge operation has to extract the real essence."[12] In order for this process to extract the real essence, it must first construct the object so that it is knowable in precisely this way, through this process. The real object has an essential part and an inessential part, the *visible surface* and the *invisible* kernel.[13] The real object is not in itself this duality, rather the duality imputed to the object is the trace of the operation of knowledge. If what is presents itself as this duality of inside and outside, essential and inessential, this is only for the process of knowledge Althusser calls empiricism. Therefore, that very duality itself is the trace of empiricism realized in the real object itself such that the process of knowledge becomes "indispensible to the discovery of the essence."[14]

The connection between this process and Aristotle's path is clear. The attempt to think that which is, particularly when it is material, turns out to find its expression in thinking only that which is thinkable in it, that is, the essence. What Althusser draws from this, however, is that this process identifies knowledge of the real object with a part of the real object itself. Knowledge is possible because the real object has been constituted in advance in such a way that the abstraction of the essence is the grasp of the object's reality. Because it is hidden, it must be uncovered, drawn out, extracted. Nevertheless, Althusser notes that knowledge in this sense cannot be something in addition to the real object, for otherwise we would forever be left with the question of the knowledge of knowledge, that is, how can one know that one's knowledge is actually *of* the real object? This was the "solution" of Aristotle's path, to make the reality, the actuality, of the thing that which is also known when the thing is known.[15] In this way, Aristotle is the founder of this kind of empiricism. Empiricism, for Althusser, "solves" this problem by adding to the real object a "*new existence,* precisely the existence of its knowledge . . . completely *inscribed in the structure of the real object,* in the form of the difference between the inessential and the essence."[16] This difference is the trace of this process of knowledge. However, Althusser points out that the knowledge of the real object is written into it and, thus, there is an attempt to eliminate this trace. In this way, Althusser uncovers the "prehistory" of the Aristotelian path. Its errancy, therefore, was necessary because the dualization of essence and dross, which originarily belongs to knowledge, is written onto and into the real thing itself. Aristotle's path is simply the reading off of the real what was inscribed into it in a silence.

As I argued in chapter 1, Aristotle wandered off the path precisely because, while he was attempting to uncover the principles of what is so that we can think it, he ended up delivering only the principle of what is thinkable. Althusser, however, uses stronger language than "errant." For him, this is a murder scene and the corpse has to be spirited away. This could be understood by returning back to his notion of trace and Derrida's explication of the strange ontology and operation of the trace. If the result of Aristotle's path, of empiricism's process, is the extraction of the essence, this is only possible by the recognition of the *difference* between the essential and inessential. Since the inessential and the essential are marked by a fundamental difference, to the extent that the essence names the reality of the real, the inessential, the dross, slips into un- or nonreality.[17] The difference, however, marks the fundamental role of essence—to be the essence *of* something. The essence depends more on the inessential than the inessential on the essence. In fact, it is only the inessential that grounds the essentiality of the essence—without something other for which the essence is essential, the essence itself proves inessential. In this way, the *difference* as such is the trace of the operation of empiricism. The essential then becomes the proper object of knowledge, it is what knowledge is *of*.[18] In going on to claim the essential as the real object of knowledge, empiricism recognizes that there is something in the real object that stands outside its grasp, and thus it does not know the *real* real object. Althusser points to Marx as the most successful investigator of this crime scene.

In the *Grundrisse,* Marx presents a terse description of his method of political economy. The problem that Marx confronts is that, on the one hand, political economy deals with that which seems to be both immediate and concrete. "It seems to be correct to begin with the real and the concrete, with the real precondition, thus to begin, in economics, with, for example, the population, which is the foundation and the subject of the entire social act of production."[19] On the other hand, population is an abstraction. That is, population as such is not entirely intelligible because it is not basic, fundamental, concrete, but rather is itself a composition—population is made up of classes. Classes, in their turn, are unintelligible because in order to understand them, we need to understand elements like wage labor and capital. There is, then, a difference that Marx sees between the way in which something appears to observation as concrete and the way in which that appearance as concrete covers over the abstractions on which it is based.

There is, therefore, a false concrete that is false because it is a "chaotic representation of the whole" [*eine chaotische Vorstellung des Ganzen*].[20] The problem, therefore, is to arrive at a correct conception of the concrete. This is a problem because the movement of thinking is not identical to the movement of the real. This was the point of departure of Althusser's critique of empiricism. "The concrete is concrete because it is the concentration of many determinations, hence unity of the diverse. It appears in the process of thinking, therefore, as a process of concentration, as a result, not as a point of departure, even though it is the point of departure for intuition [*Anschauung*] and representation [*Vorstellung*]."[21] Marx clearly recognizes that thought operates according to a logic that is other than the mode by which reality is constituted.[22] The challenge of method, therefore, is to bring thought to conceive of the real, all the while recognizing that the process of thought is not identical to real processes.

If Marx is pointing out that what presents itself in intuition does not, at the same time, present its reality, its real essence, then he is more or less an empiricist in Althusser's sense. Althusser, however, brings us to this discussion of method as a counter to empiricism. Yet on his reading, Marx is here opposing the role of thought to the real that is immediately given.[23] That is, Althusser reads the move to abstraction precisely as the method of classical political economy and, therefore, empiricism. It is difficult to reconcile that reading, however, with Marx's own move from the chaotic whole that is presented to intuition and representation to the "ever thinner abstractions" that yield the determinations on the basis of which the real concrete would be thought as determinate and not as chaotic. Althusser's blunt opposition between thought and the real misses what I argue is the most important point in Marx's method.

Marx sees the classical political economists as grounding higher-level, complex concepts, categories, and operations in a reality that they must assume is immediately available to sensation and intuition. If we take even a seemingly simple form of production like farming, we need to take account of the population. What could be more basic for economics and more obvious to observation than the people of a given nation-state? From here, we can build up relatively simple formulae of supply, demands, consumption, and, therefore, value. The difficulty with this method is that it takes population as given simply *as* population. In other words, it takes population as immediate and, therefore, in no need of analysis. The classical method takes what is immediately given to intuition as if it wears its mode of givenness on its face. However, *how* a given population is constituted is crucial for the higher-level categories, concepts, and processes. Classical political economists assume that population is a starting point and not a result. Marx's contention is that if something like population is the result of a real process, then *how* it comes to be what it is becomes central to even its immediate presentation to intuition. The gesture of classical political economy, however, points away from the *how*, from the possibility of mediation. The magic is produced, therefore, by misdirection. If we attend to how this people became the people they are, then certain forms of production, for example, become less "obvious," less immediate.

This method, the "scientifically correct method" [*die wissenschaftlich richtige Methode*], begins from the chaotic representation of the whole and then, "by means of further determination, move[s] analytically towards ever more simple concepts, from the imagined concrete toward ever thinner abstractions until [it] arrived at the simplest determinations."[24] We see, therefore, what made the original representation chaotic. It lacked, at least initially, determination [*Bestimmung*].[25] It cannot be the case that, as real and as presented to intuition, "population" is indeterminate. Rather, its determinations are not presented immediately to thought. The determinations that thought arrives at are, as it were, too late for the concrete given in reality. What immediately presents itself as concrete is not a result *of* thought, but can be thought only *as* a result. This is the core of Marx's materialist method: the real is determined by that which does not present itself immediately. Thought is required in order to prevent thought from claiming that it is determining of the real. The fact that thought might take itself as determining the real is just the other side of the process by which what is thinkable is the reality of the object and everything other than what is thinkable is "dross."

While Marx argues here that the process of thought is the reverse of the process by which the concrete comes into being, there is no need to take this strictly. That is, we need not necessarily maintain that the two processes are the same, though working in opposite directions. Rather, the basic insight of Marx is that the concrete itself is not a result of the determinations of thought. That is, the concrete itself has principles and a "logic" that is other than the principles and logic of thought. According to Marx, Hegel failed to recognize just this otherness of thought and what we might call "the real."[26]

"In this way, Hegel fell into the illusion of conceiving the real as the product of thought concentrating itself, whereas the method of rising from the abstract to the concrete is the only way in which thought appropriates [*anzueignen*] the concrete, reproduces it as the concrete in the mind [*als ein geistig Konkretes*]."[27] (Marx, n.d., 42:101; Marx 1993, 632)

Here, Marx uses language that is strikingly similar to his description of the capitalist process of production. In order for surplus-value [*Mehrwert*] to be turned into capital, the capitalist must appropriate [*anzueigen*] a portion of the product of others' labor, and must do so "without equivalent."[28] Marx raises two interrelated concepts: equivalence and appropriation. As far as the first, equivalence, is concerned, Marx exposed, already at the outset of *Capital*, the concept of exchange-value precisely as an equivalence, "1 quarter of corn=x cwt. of iron."[29] However, this equivalence is not simply mathematical. Marx's entire argument hinges on the fact that the equivalence is an identity.[30] Through some third, the corn is taken as identical to the iron. In order to be identical, however, the commodities must be abstracted [*abstrahieren*] "from the material constituents and forms which make it a use-value."[31] The identity is brought about precisely through this abstraction from the dross of the matter and all those qualities that make an object suitable for satisfying needs. Marx shows that what makes an object useful is its accidental, material characteristics and, therefore, these must be cast aside in order for two commodities to be identical. This identity is what makes exchange-value emerge.

This process, as might already be obvious, is the social form of what Althusser named "empiricism." The result of this process, according to Marx, is that "[a]ll its sensuous characteristics have been extinguished."[32] The thing, abstracted from its sensuous, material characteristics now is only a "ghostlike" [*gespenstige*] objectivity—an objectivity [*Gegenständlichkeit*] that haunts the space between the real and thought. Identity is not a given, is not found in things, but is the result of this abstraction. The phenomenon described here is the errant path of Aristotle's *Metaphysics* now traced socially.

The second term Marx raises here, appropriation [*Aneignung*], functions both in the capitalist mode of production and the "scientifically correct method." The capitalist makes its own the product that originally is the property [*Eigentum*] of labor. In this mode of production, the act of making one's own, of appropriation, occurs "without equivalent" in that labor is not returned something for that which has been expropriated. The very possibility of appropriation emerges on the basis of the abstraction from the material and sensuous characteristics of the commodity. The equivalency between commodities, the ghostlike objectivity of exchange value, is the condition for the possibility of its appropriation without equivalent. To make one's own [*eigen machen*] is to recognize, at the same time, a nonidentity, a nonequivalence. If there were an identity—either between two commodities or between thought and the real—then no process would be required in order for the equivalence to show itself. That there must be abstraction, that there must be appropriation—with regard to commodities and to the real—is itself the trace of the nonidentical, of this fundamental difference between two commodities and between thought and the real. In each case, the difference is marked by the sensuous, material characteristics of the object.

In both cases—that is, in the capitalist mode of production and the method of philosophical materialism—the equivalence is a shadow, a phantom, a specter.[33] In the Introduction, Marx marks this ghost in his recognition that thought is a process of production, of re-production, that, just as the production of commodities requires abstraction in order for the thing to become appropriate for, the property of, thought. Appropriation marks the nonidentity of thought and the concrete. The real concrete and the concrete-in-thought are each appropriate to themselves and thus are different.

The capitalist mode of production makes identical two nonidentical things by means of an equivalency that arises only after the sensuous, material characteristics, the dross, has been cast off, left behind, made inessential. What is left is a nonmaterial, nonsensuous substance. As disembodied, exchange-value is then free to roam, to haunt, and ultimately to be appropriated by the capitalist without equivalency. Yet the disembodied substance is no less effective for having no matter and no sensuous characteristics. Marx's methodological project is to uncover the ghosts and specters that "also form a part of reality."[34] If, for example, real objects wore their reality on their face, exchange-value would be impossible. Since concrete, material effects arise from nonmaterial and nonsensuous substances operate in the real, a materialist philosophy must turn toward concepts that are appropriate to such a haunted world.

As a result, Marx insists that the proper materialist method cannot take sensuous, material things at face value. To do so would not only blind philosophy to the effectivity of nonmaterial substances that have a ghostlike objectivity but also would accept only a "chaotic representation of the whole." To assume an identity between thought and thing is to appropriate without equivalence. Such appropriation comes in two forms: (1) as the idealist claim that thought produces the real and (2) as the empirico-materialist claim that thought is just the recording and reflection of the sensuous, material object. Marx's recognition that the concrete-in-thought is the real concrete made thought's own is, at the same time, an insistence on the irreducible nonidentity between thought and the real.

The recognition of this difference means that, if one is to think the real, one should attempt to appropriate it in and for thought, make it thought's own, and reproduce it as something thought [*geistig*]. This appropriation and reproduction, however, is only possible if the difference between the real and what is thought is maintained. It is, however, maintained even when negated. It is maintained by Aristotle in his differentiation of the essential from the "dross." It is maintained, as Althusser argued, by the same gesture in empiricism. For Marx, the insistence that the real be appropriated is at the same time an insistence that the real is not identical to thought nor its product. If the determinations of thought were the process by which the real itself came to be, no appropriation, no reproduction would be necessary.

There are, therefore, two dangers. The first danger is the assumption of a "false concrete," for example population, as real. This danger leads to the evaporation of the complete representation[35] and the result is an abstract determination, without the more basic concepts that determine the concrete. The second danger, however, is the assumption that the determinations arrived at in thought are taken as the principles by which the real comes into being. These two dangers, however, share a common element: they both refuse the difference between thought and the real. The first danger refuses the difference by assuming that the real, the concrete, presents itself immediately to intuition or perception [*Anschauung*]. That there is no mediation involved must also mean that there is no (essential) difference between the real and its intuition. As such, the intuited forms the ground of a deduction.[36] From population we can deduce division of labor, money, and value. From these, we can ascend to state, exchange between nations, and the world market.[37] Here, the problem is that the apparent concrete, population, is nothing at all without the apparent abstractions of division of labor, class, and so on. If I assume it as given, then the move to states, exchange between them, and the world market will prove not to appropriate the real into thought and reproduce it, but rather will falsify the concrete from which we began. The second danger assumes that whatever is real and concrete is the product of thought, and therefore has come to be according to the same principles by which it has come to be thought. This movement, which Marx ascribes to Hegel, misses the very point of its own departure: it wanted to think something.

The method Marx outlines here, therefore, is one that reproduces the real concrete in thought and makes it its own without ever making the metaphysical assumption that the principles and movement of thought are the principles and movement of reality. But it also refuses the opposite metaphysical gesture of assuming that the difference between thought and the real makes no difference, that thought can adequately capture the real as such. The assumption that the principles and movement of thought are the principles and movement of the real, that is, the assumption that the real is a *product* of thought, is idealism. The assumption that thought adequately captures the real as such is a naïve realism that leaves the concrete shrouded in the mystery of its givenness. This sort of realism can also be the foundation of a materialism that says whatever is must be matter, as if the givenness of matter requires no further analysis. Marx's method, in contrast to these two, seeks to reproduce the real in thought and make it its own for the sake of an analysis of the real, while recognizing that the concrete in thought is other than the real concrete because they come to be through different determinations.

Marx provides an illustration, returning to the categories of political economy.

> For example, the simplest economic category, say e.g. exchange value, presupposes population, moreover a population producing in specific relations; as well as a certain kind of family, or commune, or state, etc. It can never exist other than as an abstract, *one-sided* relation within an already given, concrete, living whole. As a category, by contrast, exchange value leads an antediluvian existence.[38]

Recall that exchange value is not something that appears immediately to intuition or perception [*Anschauung*], but is the result of thought moving from what is given to the abstractions that determine it. Now we see that this category does not have either an ontological or logical independence, but rather is wrapped up with population, class relations, and social formations. The abstraction discovered in thought, exchange value, has a mode of existence within a given concrete whole.[39] The category of exchange value, as opposed to its real function, is always an abstraction, yet it is an abstraction that thinking requires in order to make the real its own. This is why it appears to emerge from the distant past, like a dinosaur. This tempts us to assume that this category that emerges for thought is productive of the social whole whose givenness led to the abstraction of the category.

In this way, Marx recognizes that the method of political economy, which is, in the end, a method of materialist philosophy, must deal directly with the most fundamental problem that the analytic categories that emerge for thought are the only way to think the real concrete and yet these categories cannot be taken as the ontological principles of the real concrete—thought can only reproduce the real concrete as its own. This is a method for philosophical materialism precisely because it acknowledges that there is something other than thought that is the "object" of thought. The obviousness of this problem is precisely the reason for its difficulty. To posit that being is matter or that whatever happens to be is material runs the risk of positing matter as it emerges *in thought,* which is not matter at all. There is something that presents itself to perception or intuition [*Anschauung*] as concrete. Yet its concreteness cannot be grasped in perception or intuition but rather requires thought to uncover its determinations. This requires concepts that are ever more basic, and even abstract, and yet, nevertheless, are required in order for thought to "rise to the level of the concrete." The mere supposition that whatever exists is matter posits matter as a chaotic whole, as, indeed, something mystical. Therefore, what appears to be materialism on its face turns out to be either theological or, at best, idealistic. It is theological if it simply posits some matter as that which is by its very nature but is unable to be completely grasped because it exceeds our rational capacities. It is idealistic if it posits matter as something that is adequately grasped by thought because either produced by or identical to it. Marx's method, on the contrary, insists on a fundamental difference between thought and what is other than thought. This insistence, however, does not risk being either theological or idealistic precisely because the *other* emerges as other, as different, as nonidentical, as a result of thought recognizing ever more determinations and thus rising to the concrete.

Marx goes one step further. It is not just that he has a preference for materialism or that he insists on materialism because it is the only philosophy that can lead to the social and political consequences he has chosen in advance. Rather, he argues here in this Introduction that materialism is necessary to avoid the contradictions inherent in idealism or naïve materialism. In the *Nachwort* to the second edition of volume one of *Capital,* Marx returns to this "materialist basis of [his] method."[40] There Marx indicates that the "method of presentation" [*Darstellungweise*] must be different from the "method of inquiry" [*Forschungweise*]. On the one hand, the method of inquiry "must take in the material in detail . . . and to track down their inner connection."[41] Once this work is accomplished, the actual movement can be accordingly presented. This work, according to Marx, is preliminary to the ideal reflection of the life of the matter.[42] Once reflected back into the ideal, it can seem as if the one is dealing with an *a priori* construction. The result appears as ground.

This is the context in which Marx distinguishes himself from Hegel. It is not that the ideal has no role to play in Marx's thought.[43] Rather, the ideal appears as a reflection of the real world and not as the principle of its being. If Marx argues that the concrete does not present its determinations except to analysis, then he must also insist that thought has a role to play in materialist philosophy. If the "object" of thought is the ideal, then materialism is opposed to idealism in a way other than refusing any role to the ideal. In the afterword to the second edition of *Capital,* Marx once again outlines his method in relation to and in opposition to that of Hegel. The affinity between the Marxian and Hegelian dialectic turns on the role of the ideal and the entire argument focuses on the moment at which the ideal appears. Once again, Marx shows that for Hegel, "the process of thinking . . . is the creator [*der Demiurg*] of the real, which only forms the external appearance of the idea." Marx's materialist method reverses this metaphysics: "the ideal is nothing other than that material translated and transferred into the human head."[44] The δημιουργός must here function as both literal and metaphorical. Plato's Demiurge and its neoplatonic variants are involved, literally, in public works projects. They are artisans of what can be public—what belongs to the *demos*—they perform the work (*ergon*) of the *demos.* Since Marx has shown that thought does its work in private, it could never be the creator, *ex nihilo,* of anything. Marx's method shows that, whatever thought produces, it produces on the basis of something irreducibly different from it. Therefore, the productivity of thought can only be a reproduction. Here again we see that the ideal is the reflection of the real, of the material, and is not to be thought as its origin.

Hegel may here be nothing more than the culmination of the essential task that metaphysics as such has always set for itself, the task that is the result of Aristotle's errant path. According to Marx,

Hegel realizes this dream in positing the ideal as productive of the real. His opposition to this positing, however, is not simply to reject the ideal, to reject the process of thinking. This would lead to a mystification in its own right. While the Hegelian mystification shrouds matter and the material world in a cloak of thinking, the mere rejection of the ideal would take the abstract (e.g., population) for the actual concrete. Marx's materialism, therefore, amounts to the critical deployment of thinking, and even the ideal, in order to "rise to the concrete." As we have seen, this method is required because what presents itself to sensation and intuition is never simple, never immediate, and never grasped in its concreteness. The task, therefore, is to analyze what presents itself to sense and to intuition so as to discover the actual concrete that is cloaked within this seeming immediacy.

The question of immediacy returns over and over again whenever there is an attempt to analyze *how* thought can think the real. Here again, we find another duality. On the one hand, there is the immediacy of the "object" to thought, that is, the claim that the real, simply and without mediation, shows or presents itself to thought. Whenever there is no mediation, there is no question, there is no difficulty for thought. This is perhaps the main reason why the immediate forms the basis of Aristotelian demonstrative science.[45] The difficulty of thinking the real is "solved" by asserting that "there is nothing to think here." However, on the other hand, the real itself is also thought as immediate, that is, as not having been mediated in its being or becoming actual. This is what Marx called a chaotic representation of the whole, namely the positing that there are no processes, no structures, and no social forms that are involved in the production of the real. This duality of immediacy—the immediacy of the being of the real and the immediacy of the real to thought—are coimplicated. Only that whose being is immediate can present itself to thought immediately. However, as Althusser argued, this coimplication of these two forms of immediacy is the very trace of the process of knowledge and not of the being of the real.

Marx unpacks this problem of immediacy by noting two complexities relating to that which presents itself to sensation and intuition. First, the presentation to sense and intuition does not carry along with it the mode of presentation itself. *How* the material world shows itself to sensation and intuition is crucial in coming to grasp in thought the actual concrete. This mode of presentation, however, is discovered not through sensation or intuition itself.

This is precisely why something appears to be concrete even though it is abstract.[46] A move to abstraction, and therefore to the ideal, is required in order to come to trace the mode by which something is given to sensation and intuition at all.

Second, what presents itself to sensation and intuition is a *result*, that is, the result of sociohistorical processes. There are two directions here. The real concrete is a result of a real process. It is, however, the beginning for thought. The process of reproduction in thought, of the appropriation of the real by thought, results in the concrete-in-thought. This again has to do with the mode of givenness. In the *Nachwort* to the second edition of *Capital*, Marx points to this notion as one where he and Hegel agree. Marx notes that Hegel's philosophy turns on the recognition that there is no being without becoming. The mystical element of Hegelian dialectic does not prevent the recognition of its "rational kernel." This rational form of the dialectic, which is contained already within Hegel, "because it includes in the positive understanding of the existing at the same time also the understanding of its negation, of its necessary demise, it grasps that developing form in the flow of movement, thus also according to its transitory side, it does not allow imposition and is in its essence critical and revolutionary."[47] Dialectic, therefore, is the only method that grasps the movement and becoming that stands behind that which exists and which presents itself to sensation and intuition. Dialectic, therefore, is the method that is always on guard for the mediation of that which seems immediate. It is, therefore, the only method suitable to thinking the materiality of matter. Traditional political economy took merely the result of this process, and thus mistook what is complex and constructed for what is simple and immediately given.

Marx's method is one, therefore, that does not insist at the outset that whatever exists is material, that is, that matter as such is being. Rather, Marx recognizes that matter is that which is other than thought, since thought must reproduce it as its own. This fundamental insight is the hinge upon which his method turns. His goal is the proper understanding of material reality. The complication rests in the fact that this understanding takes place *in thought*. Because of this fundamental insight, materialist metaphysics turns almost entirely on method. If matter is that which is other than thought, and yet is grasped *by thought*, a method must be pursued that allows matter to emerge as a result of thinking and not as a simple presupposition of it.

The assumption that there is something other than thought arises in two ways. First, it arises because of the demand of analysis. As we have seen, Marx argues that what appears to sense and intuition is never itself immediate and the real concrete. As such, thought must pursue an analysis so that the actual concrete can emerge from what is given to sensation and intuition. Second, that there is something other than thought emerges directly from thought itself. Marx points to this when he refers to the mystificatory side of Hegelian dialectic. The Hegelian move to posit the ideal as the principle of the real fails just because it does not make good on the promise of thought as such.

Marx's argument, however, is that as long as our concrete, material world is haunted by ghostlike objectivities, we cannot refuse the conceptual abstractions of thought because otherwise we would not be able to think the real.[48] Exchange, as Marx argued, requires an equivalency that stands outside the material, sensuous characteristics of what is. That equivalency, the identity of two things that are not identical, produces the ghostlike objectivity of exchange value. In this way, there is the production of a false truth. "Exchange as a process has real objectivity and is untrue at the same time, transgressing against its own principle, the principle of similarity."[49] This nonidentity at the heart of identity is what calls for Marx's method of materialist philosophy—a method that Adorno calls "negative dialectics," which, therefore, "is tied to the highest categories of identity-philosophy" (ND 150/147).[50]

Marx uncovered the way in which these ghostlike objectivities are produced as part of a social process. As a result of a social process, they are *real*, that is, they are effective in the world in which we find ourselves. It is because of the production of identity that, for Marx, we cannot give up on the analytic power of thought, even though thought can easily turn these ghosts into actually existing things. For Adorno, this means that thought must be dialectical:

> In fact, dialectics is neither only a method nor something real [*ein Reales*], naively understood. Not a method: for the unreconciled matter [*Sache*], which lacks exactly that identity that thought substitutes, is contradictory and balks at any attempt at its unanimous interpretation. Matter, not the organizational drive of thought, brings about dialectics.

The identity of concept and what is, the factical, is the metaphysical form of the principle that underlies exchange. For Adorno, there are, as it were, three sides to this identity, the identity that Althusser labeled "empiricism." First, there is the principle of exchange in which two commodities are equal and, therefore, identical only by stripping away all the material and sensuous characteristics that make the thing, the *this*, what it is as given. Second, since this identity is possible only by means of a third—corn and iron are only identical because they are equivalent to something outside—and that third is the "abstract universal concept of average working hours."[51] Last, is the "principle of identification," that is, the positing of what is thought in a concept (the essence) as the reality of what is given. These three sides are, for Adorno, inextricably connected in such a way that the critique of any one must necessarily involve the critique of the others.

The identification of the factical with the concept is therefore just one side of the problem of thinking that which has been stripped away in the process of knowledge. The intertwining of all three sides of this identification—the commodity, the fungibility of every individual, and the identity of concept and reality—forces philosophy, according to Adorno, to remain tied to the ultimate categories of identity-philosophy. What has been left behind as "dross," the material and the sensuous, that is, that which makes this thing be this thing, cannot be thought immediately and directly. "The force that shatters the appearance of identity is the force of thinking: the application of the 'it is' of identity shakes the form of that appearance, which remains inalienable just the same" (ND 152/149).

The methodological problem for which negative dialectics is the solution emerges once again because of the reality of that which is not existing, that is, from the ghostlike objectivity of concepts haunting reality. If the world were right, then it would be impossible for the identity of concept and thing to seem as if it were true. Again we face the errant path of Aristotle's philosophy, the path that began with the attempt to think that which is and ended up thinking only the essence, that is, only what is thinkable in it. Adorno, following Marx, insists that the error does not lie in thinking the essence, which is all that abstract concepts can do, but in thinking that the essence is the reality of that which is.

> To think is, according to its own [*eigen*] meaning, to think of something. Even in the logical form of abstraction of something, as [something] meant or judged, which of itself claims to expose no being, lives irredeemably that which thinking would like to redeem, whose non-identity is that which is not thinking. *Ratio* becomes irrational where it forgets this, where it hypostasizes its products, the abstractions, against the meaning of thinking.[52]

The Aristotelian "this something," already includes an abstraction. Yet within that abstraction lives that from which thought has abstracted. It is a debt that thinking tries to pay off by means of a replacement: instead of thinking this thing that is, thought replaces the essence to pay off the debt.

While we are hesitant to think of ghosts as forming a part of reality, Adorno is here insisting that just such a spectral form of existence, that is, debt, is not only real, but comes back to haunt thought. If thought is to think some thing, it can only do so by means of the abstraction from the given thing to "something." This move from this thing here in front of me to "something," is certainly valid, logically. From "an X," to "some X," to "something," is, from a logical point of view, equivalent—it is not even a deduction. However, if thought moves into this logical equivalency, it is because it is always of "something." There is no thought as such, no empty thought. Therefore, "something" is an abstraction away from the factical but *in the name of* the factical. The abstractive demands of thought always presuppose a given from which thought abstracts. In this way, thought is only productive on the basis of a transposition of an absence—the factical—into a presence—the concept is the reality of the factical.

For Adorno, this attempt to pay off what is owed is tantamount to paying off a debt with a promissory note. While the very meaning [*Sinn*] of thought is to think *something,* the moment it insists that the concept is the reality of what it is, it shows that it only thinks what is thinkable. Here it falls short of what it is in its own meaning. Thought comes up short and, therefore, has something to make up, to make good on, to redeem. Aristotle, and, indeed, metaphysics from Aristotle through Hegel, redeem the debt through a kind of bond, that is, another debt: since Aristotle failed to think that which is, he paid of that failure through the security of essence as the reality of what is. There is a fault, a failure that belongs not to the factical, to what is, but to thought. Thought is parasitic on what is and therefore always owes something to the given, the existent, the factical. Yet Aristotle's errant path is nothing but the attempt to pay off that debt with an abstraction. Even in this abstraction, thought still recognizes the other to which it owes something and, therefore, that other lives on, even if only as that to which thought is in debt. This "living on," is that which in the thing is not identical to what thought can think. Nonidentity is the basis of the identity-philosophy that begins with Aristotle.

In this "living on," matter emerges for thought, but it emerges as difference from thought, as the other than thought. If thought is to think "that which is," that is, the factical, then it must think it in this otherness, in the nonidentity of concept and thing. That Adorno places this relation in the economic one of debt and payoff shows that, for him, the errancy of Aristotle's path is not a "mistake," but rather just what happens when the concepts of thought are identified with what is. A debt is incurred because thought was supposed to think that which is and thought attempts to make good on this by turning what is thinkable—the essence—into what is real in the thing.

It is the notion that what is other than thought lives on even in thought that brings Adorno to materialism. This is not a direct connection, precisely because there is no immediate access to what is not identical to thought. First, Adorno recognizes what he calls the "priority [*Vorrang*] of the object" (ND 184/183). The object has a twofold priority over the thinking subject. First, every subject, every "I think," is at the same time an object. Therefore, objectivity, being a thing, belongs to the very definition of a subject, while being a subject does not belong to the definition of an object. There is no necessary relation between any given thing and the thinking of that thing, yet every thinking is a thinking of something, of some object. This is the second priority of the object.

The priority of the object, of the factical, however, does not make of it a starting point. We will come to see that this version of materialism comes to be associated, in Adorno, with theology. Rather, even the claim of Descartes' *Meditations* that I *am* includes a claim about the objectivity of the thing that thinks. The "is" is not simply an "is thought" but indicates a being that is not dependent on thought. Yet this being that is not dependent is not established by the subject, by thought. The object, therefore, will always be constituted by the element of otherness, even when that object is the "I" that thinks. There is, however, a difference in the otherness that is supposed to be the object of thought and the otherness that belongs to the subject when it thinks "I am." The otherness of the subject is always mediated by the otherness of the object. That mediation is key to the possibility of recognizing that which is not identical to thought. "Only because the subject, for its part, is mediated, and also not the radical other of the object, which first legitimates it, can the subject grasp objectivity at all" (ND 186/185). The mediation of the subject through objectivity becomes the key through which thought can come to open onto that otherness, that which is other, that which is not identical to it.

Because the subject is not immediate, even to itself, its mediation can serve as the path to recognize the mediation of the object. Immediacy is no solution to this problem. The difficulty is, as we saw with Marx, that what presents itself immediately is not, for that reason, concrete, that is, immediate. "The dialectical mediation of the universal and the particular does not permit a theory that opts for the particular to overzealously treat the universal as a soap bubble" (T. Adorno 1970, 6:200; T. Adorno 1973, 199). This is precisely what forces dialectics on philosophy. There is mediation on all sides. That which is is mediated, the thought of that which is is mediated, the relation between thought and what is is mediated. In each case, there is mediation because there is an otherness, a difference, a nonidentity that is lost in equivalence and identification. It is this need for dialectics that is, at the same time, a need to think matter, to think materiality, by means of the concepts, abstract as they are, of thought. On the one hand, those concepts are not mere "soap bubbles." On the other, they are not identical to what is. This is what led Marx to rethink what it means for philosophy to be materialist. It is what led Adorno to a dialectics that is negative and immanent. There is a task, a power, that concepts have that is required in order to think the materiality of what is as other than thought.

NOTES

1. This situation—"it cannot be . . . it must be"—is precisely that which opens a dialectic. Aristotle insists frequently that matter must be a principle of what is while his critique of Thales argues it cannot be. For a possible resolution of this question that is not dialectical (in the modern sense), see Kirkland, "Stupsnasigkeit. Oder: In Errinerung an Aristoteles und die Form der Materie." *Erinnerung* (2012): 21.

2. I leave aside here the question of whether everything that is must also be material. We return to this in chapter 4, where we see Hobbes's argument that being is just to be a body and to be a body is to be material.

3. Louis Althusser, *Reading Capital* (New York: Verso, 2006), 35; Louis Althusser, *Lire Le Capital* (Paris: François Maspero, 1973), 38.

4. Althusser, *Reading Capital*, 35; Althusser, *Lire Le Capital*, 39.

5. This is also, I would argue, the main concern with which Descartes begins the *Meditations*. There, he announces that the difficulty is accepting as true what has turned out to be false. That is, he has accepted as what is (taking Aristotle's relation of the true to "what is") something that turns out not to be. If we take Descartes seriously, that is, if we accept the *Meditations* as meditations on *First*

Philosophy, then his problem turns out not to be epistemological but the more fundamental relation between the true and being.

6. Althusser, *Reading Capital*, 35–36; Althusser, *Lire Le Capital*, 39.
7. Althusser, *Reading Capital*, 36; Althusser, *Lire Le Capital*, 40.
8. Althusser, *Reading Capital*, 36; Althusser, *Lire Le Capital*, 40.
9. Althusser, *Reading Capital*, 36; Althusser, *Lire Le Capital*, 40.
10. Jacques Derrida, *Speech and Phenomena: And Other Essays on Husserl's Theory of Signs* (Evanston, IL: Northwestern University Press, 1973), 156.
11. Ibid.
12. Althusser, *Reading Capital*, 37; Althusser, *Lire Le Capital*, 41.
13. Althusser, *Reading Capital*, 37; Althusser, *Lire Le Capital*, 41.
14. Althusser, *Reading Capital*, 37; Althusser, *Lire Le Capital*, 41.
15. This is what Adorno calls "identity thinking." We turn to this concept in the final section of this chapter.
16. Althusser, *Reading Capital*, 38; Althusser, *Lire Le Capital*, 42.
17. Adorno claims that this is the conclusion of Plato's philosophy, namely, that the factical becomes nonbeing. See Adorno, *Metaphysik: Begriff Und Probleme*, vol. 1, Nachgelassene Schriften: Vorlesungen (Frankfurt: Suhrkamp, 1998), 1:28–34; Adorno, *Metaphysics: Concept and Problems* ed. Rolf Tiedmann, trans. Edmund Jephcott (Stanford: Stanford University Press, 2001), 15–19.
18. I have to insist here on a certain "Chicagoism." In Chicago, it is not unusual to hear a sentence like, "Would you like to come with?" After Dryden, the ban on ending sentences with a preposition (almost certainly based in Latin grammar and not English) became commonplace, except in places such as Chicago. With this ban, English lost a whole host of rich expressions, indeed even phrasal verbs. This, however, was not without a trace. For example, many "vulgar" expressions rely on sentence ending prepositions—"That is screwed up"—in relation to which no one awaits an object. However, here the insistence is not on the richness of Chicagoese, but rather on the philosophical point. "Thinking of" is not a phrasal verb, but *always* demands an object. Built into this is the very difference between thought and that which it is *of*. The moment that Aristotle or empiricism identify the object of thought as that which is thinkable, they not only lose the object but also turn thinking into something other than itself. Yet as Althusser has been noting and as we shall see over and over again, this is not done without a trace.
19. Karl Marx, *Grundrisse*, vol. 42, Marx Engels Werke (Berlin: Dietz Verlag, 1953), 42:100; Karl Marx, *Grundrisse: Foundations of the Critique of Political Economy (rough Draft)*, trans. Martin Nicolaus (Penguin, 1993), 631.
20. Ibid. I translate *Vorstellung* as "representation" in order to keep in line with the Hegelian terminology of this section of the Introduction. Marx clearly has in mind here an engagement with Hegel, following him to a certain extent, turning away from him at crucial points.
21. Marx, *Grundrisse*, 1953, 42:101; Marx, *Grundrisse*, 1993, 632.
22. I return below to a discussion of this in relation to Althusser. There, I argue that Althusser is a bit one-sided in his reading of Marx to the extent that he leaves unanswered the question of the relation of thought to materiality. Certainly, Althusser would see my reading here as still too Hegelian.
23. "The theoretical problematic underlying this text allows us to distinguish Marxist philosophy from every speculative or empiricist philosophy. The decisive point of Marx's thesis concerns the principle distinguishing between the *real* and *thought*" Althusser, *Reading Capital* (New York: Verso, 2006): 86–87; Althusser, *Lire el Capital* (Paris: François Maspero, 1973): 106. Althusser goes on to suggest that Marx takes over, uncritically, the categories of bourgeois politi-

cal economy and remains silent about the appropriateness of their categories. My disagreement with his reading should be clear as I proceed.

24. Marx, *Grundrisse*, n.d., 42:100; Marx, *Grundrisse*, 1993, 631.

25. As becomes clear below, I disagree with some of the specifics of Althusser's various readings of this passage. However, Althusser is another crucial moment in the attempt to think of matter.

26. While this term is central for Deleuze, Marx himself uses it in this Introduction. Marx is attempting to think through the difficulty that thought regards something other than itself. This is also what Althusser called the "real object." As should be clear, the use of the term "object" can be misleading when it is understood over against a subject. Therefore, I prefer to use it only when necessary.

27. Marx, *Grundrisse*, n.d., 42:101; Marx, *Grundrisse*, 1993, 632.

28. Marx, *Das Kapital*, Vol. 23, *Karl Marx Friedrich Engels Werke* (Berlin: Dietz Verlag, 1968), 23:609. Marx, *Capital*, 729.

29. Karl Marx, *Das Kapital*: 23:51; Karl Marx, *Capital*, trans. Ben Fowkes (Middlesex: Penguin Books, 1976), 127.

30. As I argue below, this is the central thesis of Adorno's *Negative Dialectics*, viz., the social form of equivalence in capitalist societies is the other side of the metaphysics of identity.

31. Marx, *Das Kapital*, 23:52; Marx, *Capital*, 128.

32. Marx, *Das Kapital*, 23:52; Marx, *Capital*, 128.

33. While Derrida's *Specters of Marx* uncovers the function of the specter primarily, or at least originally, from the *Communist Manifesto*, he does recognize just this function of the specter. See Jacques Derrida, *Specters of Marx: The State of the Debt, the Work of Mourning, and the New International*, trans. Peggy Kamuf (New York: Routledge, 1994), 125–76; Jacques Derrida, *Spectres de Marx* (Paris: Éditions Galilée, 1993), 201–79. In this chapter, Derrida develops a line similar to the one being developed here, though he takes *The German Ideology* as his point of departure.

34. The phrase comes from Hegel, *Vorlesungen Über Die Ästhetik* (Frankfurt am Main: Surhkamp, 1970): III:530; Hegel, *The Philosophy of Fine Art*, trans. F. P. B. Osmaston (London: G. Bell and Sons, 1920): IV:304. Hegel raises it in relation to comedy as the end—in the sense of fullness and completion—of art. His argument is that contradictions that is, the false, is not opposed to reality but has an effectiveness that the comic persona uniquely understands and masters.

35. As indicated above, this is a translation of *Vorstellung*. While indeed this may have the sense of a presentation in the mind of that which was encountered sensibly, and thus a re-presentation of it, there is no sense of a repetition, of a presenting again. Rather, it may be understood as a presentation in thought of that which was not initially an object of thought. In this way, Marx's discussion here might not fall to Deleuze's criticism of the representational "image of thought" in *Difference and Repetition*.

36. This assumption that Marx uncovers in classical political economy has its precursor in Aristotle, for whom *nous*, a knowledge of that which is without middle, forms the ground of demonstrative syllogism. See *Posterior Analytics*, trans. Hippocrates Apostle (Grinell, IA: Peripatetic Press, 1981), II, c.19.

37. Marx, *Grundrisse*, 1953, 42:100–01; Marx, *Grundrisse*, 1993, 632.

38. *Grundrisse* 101/632. The emphasis is in the German but omitted in the English translation.

39. In the following chapter, we see that Duns Scotus provides a model for thinking exactly this. However, it requires a kind of expanded ontology in

which the real is not identical to the actual or existing. This was not an insight original to Hegel. Scotus, in turn, most probably discovered this from Avicenna.

40. Marx, *Das Kapital*, 23:25; Marx, *Capital*, 100.

41. Ibid., 102/27.

42. Ibid.

43. As should be clear, and is made more clear below, this is the point at which I depart from Althusser's reading of these texts and the issues involved.

44. Ibid. Translation modified.

45. On this issue, see Richard A. Lee and Christopher Long, "Nous and Logos in Aristotle," *Freiburger Zeitschrift für Philosophie und Theologie*. 54, no. 3 (2007): 348.

46. This is the issue for which I turn to Hobbes in chapter 4 below.

47. Marx, *Das Kapital*, 28; Marx, *Capital*,100. Translation modified.

48. In the next chapter, we see that Duns Scotus develops the resources for thinking the real even under those conditions in which the real is a ghost-like objectivity.

49. Theodor Adorno, *Negative Dialektik: Jargon Der Eigentlichkeit*, vol. 6, Gesammelte Schriften (Frankfurt am Main: Suhrkamp, 1970), 6:190; Theodor Adorno, *Negative Dialectics*, trans. E. B. Aston (New York: Continuum, 1973), 190. Translation modified.

50. Adorno, *Negative Dialektik*, 6:150; Adorno, *Negative Dialectics*, 147. Hereafter this text will be referred to as ND, followed by German and English page numbers.

51. ND 149/146. In *Capital*, Marx shows that the "third" underlying the identity of two different commodities cannot, itself, be able to be differentiated into different characteristics. Therefore, the labor that underlies the exchange is universalized first in the form of quantity—average hours—and second in the form of "labor in general" or abstract labor. For Marx, this means that the equivalency in exchange is twofold: concrete labor, the basis of equivalence, is the form of manifestation of its opposite, abstract human labor. See (Marx 1968, 23:73; Marx 1976, 150). Adorno will come to point to this last point as leading to the absolute fungibility of every individual. I return to this dimension of the problem in chapter 5.

52. ND 44/34. My translation differs from that of both Ashton and Redmond. Adorno here plays on an economic metaphor: *untilgbar/tilgen*. I have translated *tilgen* as "redeem." This is not without its own problems, as the term *Erlösung* and its cognates is quite central, even if not frequent, in Adorno's work. One crucial use is from the end of *Minima Moralia*: "The only philosophy which can be responsibly practised in face of despair is the attempt to contemplate all things as they would present themselves from the standpoint of redemption. Knowledge has no light but that shed on the world by redemption: all else is reconstruction, mere technique" (T. Adorno 1951, 4:283; T. W. Adorno 2005, 247). In the present passage, the play is on debt and payoff. The factical, existing thing carries on in thought as a debt that thought would like to dispense with. I return to the other notion of redemption in chapter 5.

THREE

Duns Scotus: Materiality as Singularity

If for Marx and Adorno, if not for Althusser, materialism cannot reject the capacities of thought, indeed for them it requires those capacities, then how is that which is other than thought to emerge *within thought*? We have seen that their refusal to give up on thought stems from their understanding that the material thing grasped immediately provides no route to the critical analysis of that thing. In fact, the material thing is itself the result of a process, it is mediated, and thought is required to grasp that mediation. There is an ontology underlying this materialist philosophy. However, it is not the ontology that says only matter is. Rather, it must be an ontology that expands the notion of the real to include what is uncovered by thought, an expansion that takes actual existence as *just one way* of being real. This ontology is precisely that thought through most thoroughly by Duns Scotus.

It may seem odd to include a discussion of this High Scholastic thinker in a discussion of materialist philosophy. In fact, he earned the nickname "The Subtle Doctor" precisely because he was prone to see or even invent such subtle entities—almost ghostlike—in order to solve problems. He seems, on the face of it, to be the opposite of a materialist. However, as I will argue, Scotus presents an ontology in which the reality of what is grasped by thought is fully acknowledged. Yet Scotus is also the philosopher who shows the irreducible singularity of actually existing things, and he calls that singularity "material being." His ontology, therefore, provides a route to exposing more fully the relation between what is available to thought and that which is other than thought.

THE CONCEPT AS "HIGHER-LEVEL ESSENTIALITY"

In his *Lectures on Metaphysics*, Adorno argues that

> [T]he sphere of metaphysics in the precise sense only comes into being where [the tension between the sphere of transcendence and the sphere of that which is merely the case] is itself the subject of philosophy, where it comes within the purview of thought. It might be said, therefore, that metaphysics arises at the point where the empirical world is taken seriously, and where its relation to the supra-sensible world, which was hitherto taken for granted, is subjected to reflection.[1]

For this reason, "Metaphysics . . . always deals with concepts. Metaphysics is the form of philosophy which takes concepts as its objects."[2] Plato has no metaphysics, however, not because he lacks concepts. Indeed, Plato's Ideas can be seen precisely as concepts, as the secularization of the gods in the form of concepts. However, Plato goes one step further and maintains, according to Adorno, that the Ideas are being itself while the beings, the sensible world, become a sphere of μὴ ὄν. There is no thought of the tension between the transcendent sphere of Ideas and the merely factical sphere of the sensible world. The thought of this tension emerges with Aristotle, for whom there is a relation between concepts and that which is the case.

Adorno's *Lectures*, much like *Negative Dialectics*, is a critique of the proclaimed overcoming of the tension between concept and the factical precisely because the concept is almost always claimed as the "higher level essentiality" of the factical. However, Adorno is not willing to do away with the concept by means of an uncritical acceptance of what is. The closing section of *Negative Dialectics*, "Meditations on Metaphysics," shows that there is a promise implicit in metaphysics, the promise that things could be otherwise, that ought not be abandoned. In fact, he refers to this as the new categorical imperative for philosophy. If metaphysics itself is, for Adorno, a critique of a conception of this essentiality, it is also, at the same time, "an attempt to rescue something which the philosopher's genius feels to be fading and vanishing."[3]

THE FREEDOM OF THE CONCEPT FROM THE FACTICAL

For medieval thinkers, any concept, any act of understanding, must have an object and that object cannot be nothing.[4] If there is a concept of a thing that is not given in actual existence, then just what is the object of this concept? One way to see this problem, according to Duns Scotus, is to look at the notion of potency. For him, potency can mean one of three things: (1) not impossible, (2) not necessary, (3) not the same as actual about the same thing.[5] Scotus calls this last sense "metaphysical potency," and, since it is opposed to actuality, would be the potency of that which is not given factically. This potency is "based on essence, which is said to be possible, and is an order of that essence to *esse*[6] as to a *terminus*."[7] If potency refers to something at all and yet cannot refer to the act since, when the potency is, the act is not, there must be "something" to which the potency refers prior to any actual existence. It refers precisely to the essence in that the essence can be conceived and yet there may be no actual thing of which it is an essence.

Scotus seems ultimately to reject this notion of potency, a potency that would seem to give a certain kind of being (*esse*) to an essence. However, the situation is not so simple. If, for example, God knows all things eternally, then there seems to be an eternal relation between God and something other than God, a relation of knowledge. The difficulty is that if a thing does not actually exist, God's knowledge would seem to have no object. The common starting point among medieval Latin thinkers was that God knows all things through the divine essence. Scotus argues that this relation, then, between divine intellect and thing known, related through the divine essence, will be a real relation and not a mere relation of reason. His arguments all hinge on the fact that any relation of reason between God and thing known will rely on a prior real relation. However, on the way toward proving this, he must think through the possibility of a concept removed from a connection to the factical.

Scotus argues that any relation of reason that comes from a comparison of the object to something else by means of the intellect is in the intellect only as a "diminished being" (*ens deminutum*). This is a lesser kind of being because it has only "*esse* in the intellect as the known in the knower."[8] The object known—and here Scotus is considering God's knowledge of the essence or quiddity of something other than God—is known through a relation, viz., the relation to the divine essence, but that relation too has only diminished being in God's intellect. This kind of *esse* is different, Scotus argues, from the essence "as it is something in itself."

Even though the concept of *this* stone or *this* human has, as a concept, diminished being, it does have existential weight—it is the case that the concept of a nonexisting stone or human is still of a human or a stone. Scotus considers the difference between God's concept of a human that will not exist and a chimera. The difference between the two, he argues, is that "to be something" is not repugnant to the human, but it is repugnant to the chimera. The reason for this is, simply, "because this is this and that is that."[9] In other words, the concept conceives what I will call a "reality" that is not exhausted by actual existence. To the reality of the chimera, actual existence is denied because it is, in Scotus's words, "formally repugnant." To the reality of the human, on the other hand, actual existence is not formally repugnant. Scotus goes on to unpack just what is conceived by such a concept. "Human is formally human, and this whether really *in re* or intelligibly in an intellect."[10] That is, the concept conceives, if there is no repugnance, something that has, to use Scotus's word, *entitas*, "beingness," or reality. I return to this notion below.

Up to this point, the discussion has centered on God's knowledge. When it comes to our knowledge, Scotus argues against Aquinas's position that the first object of our knowledge is the quiddity of a material thing. Aquinas, more or less following Aristotle, had argued that our intellects abstract the nature, essence, or quiddity of a thing from its image in imagination—an image that arises from sensation. As such, what our intellect knows primarily and first of all is this quiddity. Scotus insists, against this position, that our intellect can know a thing under various concepts. Given this stone, I can know it as this, as a stone, as a body, and as a being. However, a given capacity cannot know something under a concept more common than the concept of its primary object. He gives the example of vision, which has color as its primary object. It cannot know the visible thing under a definition and concept more common than that of color. If it were able, then it would seem that vision is not a singular sense specified by its object. However, we can, and do, know a given material thing under the concept of being.

It is certainly the case that in this life our concepts are caused by sensitive contact with material things. The origin of our concept in sense does not, Scotus argues, entail that what is conceived primarily is that material object that was originally in sense. The intellect is capable of conceiving something more common than what is given it in sense by the material thing. The sensible thing includes, essentially or virtually, many things, all the way to being as such, and therefore being is the object that is adequate to our cognitive potency. This is how Scotus understands Avicenna's statement that being is immediately impressed on our intellect.

That Scotus argues that being as such is the adequate object of our intellect indicates that, for him, the intellect is able to have a concept of something that is beyond what is merely given, what is merely factical. This argument is precisely what pushes him to develop his famous position that being is predicated univocally of everything that is. Scotus is responding to the question of whether God is naturally knowable by us through a concept that is proper to God. His simple answer is that for us, in this life, no such concept is possible. Yet on the way to making his argument, he shows that there is a concept of God that we can have by means of the concept of being. This entails, however, that being is a concept that is univocal to God and creatures.

It should be noted, first of all, that Scotus does not reject the common understanding, found already in Aristotle's *De Anima*, that, since our intellects are the intellective souls of bodies, we can have no concept unless our intellect is moved by sense and imagination. However, Scotus argues further that

> Our intellect can abstract an object included in something immediately moving [our intellect], and consider that which has been abstracted not considering that from which it abstracts, and by considering that which has thus been abstracted, it considers something common to sensibles and insensibles, and in that it considers the insensible in a universal, just as the sensible.[11]

Since the intellect can not only draw out from what has been sensed that which has not been sensed but also can consider it without relating it back to that from which it has been abstracted, the intellect can have a concept of that which, though arising from sensation, is not tied to what has been experienced in sensation. The intellect does need a first object that moves it but that object need not relate a phantasm, an image caused by sense, to intellect.

In this way, Scotus indicates that the intellect can have concepts that are, in a certain way, freed from the origin they have in the material things present to sense. That is, the concept, for Scotus, can extend beyond the factical, beyond what is merely given in sense. Although, however, the concept has this kind of freedom, it still, for Scotus, remains tied to the factical:

> For Socrates insofar as Socrates is a singular, and nevertheless from Socrates many of the previously mentioned [concepts] can be abstracted, and therefore the singularity of something does not impede that, from that which is singular, some common concept can be abstracted. And although whatever is *in re* would be a singular from itself in existing, such that nothing contracts another there to singularity, nevertheless, that can be conceived as this *in re* or in a certain way indistinctly, and thus either as a singular or as [something] common.[12]

The "freedom" that belongs to the concept arises from this act of abstracting, of conceiving of something "in some way indistinctly." However, the freedom does not, as it were, disassociate the concept from that of which it is a concept. As Scotus indicates here and as we shall see below, to conceive of Socrates as Socrates and to conceive of him as human is *to conceive one and the same thing*. The concept human is still a concept of Socrates, but conceives of him in some way indistinctly.

However, this way of understanding should not be taken as "confused." For, as Scotus argues, "nothing is conceived distinctly unless when all are conceived that are included in the essential account [*ratio*] of it."[13] This means that if I were to give an account of the essence of Socrates, I would indicate that he is a rational animal. Therefore, in order to conceive Socrates distinctly, I must have concepts that are achieved precisely by the act of abstraction indicated previously. So while the factical singular is the first object moving the intellect, the intellect only distinctly conceives of the factical by conceiving everything included in its essential concept or account.

It is this position that leads Scotus to argue that being [*ens*] is the first adequate object of the intellect. If the distinct concept of anything is that which includes everything included in its essential account, then being must be conceived in order for anything to be conceived, since "being is included quidditatively in every inferior, quidditative concept."[14] The corollary to this is that in order for anything to be conceived distinctly, being must be conceived. And being is always conceived distinctly, since it has a "simply simple" concept.[15] Since the concept of being belongs quidditatively to the concept of any other thing and since it is a simply simple concept, being can be conceived distinctly without any of the others to which it belongs quidditatively. The conclusion is that "being is the first, distinctly conceivable concept."[16]

The argument that the concept of being is the first, distinctly conceivable concept and, as a result, the first adequate object of our intellect, differs from the position of, for example, Thomas Aquinas, who argued that, for us in this life,[17] the adequate object of the intellect is the essence of the thing that is given first in sensation. Scotus offers several arguments against Aquinas's position, but the primary focus of his interest is precisely in the ability of our intellect in this life to have a distinct concept of being that is its adequate object. First, Scotus notes what it means for something to be the "adequate" object of the intellect: "No potency can know some object under an account more common than would be the account of its first object . . . because then that account of the first object would not be adequate."[18] Here Scotus defines a notion of adequacy that entails that something is the adequate object of the intellect only when the intellect could not know that thing under a more common account, definition, or concept. To be an adequate object is just to be that which is conceived such that it cannot be conceived under a more common concept.

This understanding of adequacy means, according to Scotus, that the essence of a material thing cannot be the adequate object of our intellect because, as we have seen, "the intellect knows something under an account more common than would be the account of a material being, because it knows something under the account of being in common."[19] Scotus here insists that metaphysics as a science demands that the concept adequate to our intellect be something "higher" than the factical. This move beyond the factical, however, does not stop with the essence of the material thing but must go all the way to being, as the concept of being—along with the concept of the ultimate, specific difference—is the only simply simple concept.

For Adorno, this move from the factical to a "higher level essentiality" was the first step toward what, in *Negative Dialectics*, he calls the *Identitätsthese*: "To think is in the proper sense to think of something. Yet in the logical form of abstraction of 'something,' as that which is meant or judged, which claims to posit from itself no being, that which is not identical, what is not thought, survives the thinking that would like to extinguish it."[20] The identification of thought and the factical requires, for Adorno, this first move of metaphysics, namely the move away from whatever in the factical does respond to thought. The second step is to identify the thinkable as what is real, that is, what is essential in the factical. Finally, identity thinking shows that anything that is not essential does not make up the reality, and therefore facticity, of that which is given. In this, Adorno's concern with identity is the same as Althusser's concern with "empiricism."

Yet Scotus, for his part, rejects that very identitarian move. He reads Aquinas as arguing that "any thing is related to *esse* as it is related to knowing."[21] Against this, he argues that "a potency and object do not need to be assimilated in the way of being, for they are related as motive and mobile, and thus they are related as dissimilars."[22] The fact that intellect is "of something," means, for Scotus, that it is related to its object. Yet, he argues, that very relation, as a relation, requires a nonidentity, a dissimilarity. If, according to him, the intellect is able to conceive an object, there must be a proportion between them. However, the very notion of proportion requires a nonidentity, since identity is the absence of proportion. As a mode of thinking, the concept must be in some way nonidentical to its object.

Scotus then concludes that either nothing is the adequate object of understanding, a position that means that we understand nothing, or that "it is necessary to posit a first, adequate [object] on account of commonness [*communitas*] in itself."[23] The commonality of the first, adequate object of the intellect is precisely Scotus's position that being is univocally predicated of everything that is. If I maintain that being is equivocal or even analogical, then there would be no first object of our intellect.[24] Recall that Scotus had already argued that nothing is conceived distinctly through a concept that is more common. This move to commonality was precisely what led him to insist that being is the first to be conceived distinctly since it cannot be conceived through a more common concept.

We have seen that Scotus's argument sets out from the position that if any concept has "essential parts," then that thing is only distinctly conceived when its essential parts are conceived. We saw that this meant that any concept that is not simply simple is understood only when that which it includes are also distinctly conceived. This allowed him to conclude that there are only two simply simple concepts—the concept of being and the concept of any ultimate difference. Since any concept of an ultimate difference will be simply simple, it will not include the concept of being. The predication of being of ultimate differences will be of another sort than the predication of being of anything else that is. While being is predicated *in quid* of any factical thing, it is not predicated *in quid* of ultimate differences.[25] That is, given any thing, if we ask "what is it," "being" will always be part of the answer. However, given an ultimate difference, for example, "rational," if we ask "what is it," we cannot answer that it is a being. While being is univocal, it is not predicated univocally *in quid* of everything because it is not predicated in this way of specific differences.[26]

Because of this difference between the predication of being of a given factical thing and the predication of being of specific difference, we can note that there is a difference between, as it were, the work that each of these does. If a concept is not simply simple, it is resolvable into other concepts. If we take a concept like "human," that is, "rational animal," we can see that if animal is capable of being specified, differentiated, by rational, then rational would be a determination of animal. So, in any concept that is not simply simple, the concept is composed from a "determinable" concept (animal) and a "determining" concept (rational). This relation of determinable to determining is, for Scotus, just like the relation of potency and act.[27]

> So too it is necessary in concepts that every not simply simple, and nevertheless one *per se,* concept be resolved into a determinable and determining concept, such that this resolution stops at simply simple concepts, evidently at a concept that is determinable only, such that it includes nothing determining, and at a determining concept that, evidently, does not include some determinable concept. The concept that is only determinable is the concept of being and that which is determining only is the concept of the ultimate difference. Therefore, these will be immediately diverse such that one includes nothing of the other.[28]

The immediate diversity of the concept of being and the concept of the ultimate difference means that the ultimate difference is of a different order from being, it does not "belong" to being. For this reason, being cannot be predicated *in quid* of the ultimate difference. However, anything that is, any given being, will be "constituted" by the ultimate difference as well as by being. The work of the difference, therefore, is required for actuality, even while its work is outside of, immediately diverse from, the potency that belongs to the concept of being.

If the ultimate differences stand outside the order of being, as it were, such that being cannot be predicated *in quid* of them, then it seems as if the univocity of being is called into question. Scotus, however, recognizes that there is a relation between difference as determining and being as determinable in that whatever is determined will be a being. So the concept of a difference does have some connection to being. To unpack this, Scotus mobilizes a complex notion of virtual inclusion. Being has a "twofold primacy" as the first object of our intellect. One primacy, seen in the concept of being, is the primacy of commonness [*communitas*]. As the most common, being is the first object of our intellect. The other primacy is that of *virtuality*. "Every *per se* intelligible either includes essentially the account of being or is contained virtually or essentially in essentially that which includes the account of being."[29] Any factically given thing includes being essentially. In addition, that factically given thing also includes the ultimate difference essentially. Therefore, "those for which being is not univocal *in quid* are included in those for which being is univocal in this way."[30]

As I hope to make clear when I turn to Scotus's discussion of singularizing difference, the position that the specific difference is "simply simple," that is, not further reducible, and that being is not predicated *in quid* of it entails that being, in its univocity, is differentiated. This is part of the meaning of Scotus's argument that the difference is "virtually included" in being. As he has argued, anything differentiated will be a being. If Scotus's thinking of determination as actuality is added to this, the result is that the less differentiated is the more "diminished." What is more, the determined being does have being predicated of it *in quid*. Differentiation, therefore, belongs to being—being simply "is" as differentiated. In this way, the multiplicity of its determinations just is what being is and this is its virtuality, for Scotus.

The univocity of being does turn away from the factical, indeed turns toward a "higher level," that is, pertaining to the essence of the factical. As Adorno points out, this turn away as a gesture of identity is required for the concept to be posited as identical to the thing of which it is a concept. Scotus, as we have seen, uses precisely the notion of intelligibility in order to argue his point. Yet at the same time, the argument leads him to insist on a diversity between the concept of being, with its inferiors, and the concept of ultimate difference. Both are quidditative in that both belong to any rational account of the factical, and yet, precisely since they both are quidditative, they must be doing separate work in relation to the factical. Difference emerges, on this account, precisely as the actualizing of the potency that belongs to the concept of being—the determining of the determinate. To this point, however, Scotus has only thought difference to the level of specific differentiation. There is still further difference at work, namely the difference making an individual and not just making a species.

THE DIFFERENCE THAT SINGULARIZES

For Scotus, every factically given thing is in some sense a composite of what he calls a "common nature" and an individuating principle. Against the position that any factical thing is "of itself" a singular, Scotus argues that the common nature or quiddity cannot be a singular from itself. He has several arguments for what I will come to call the "reality" of the common nature, but one in particular is drawn from arguments we have already seen. That the intellect can abstract from any given thing a concept that is more common than the thing, and yet is the concept of that thing, means that, through a more common concept, the intellect can understand the factical, which is itself a singular. However, if that common concept is universal—or at least not singular—and at the same time conceives the singular, "the intellect, understanding that object under a universal account, would understand it under an account opposed to [singularity]."[31] Since the intellect can abstract from the singular a concept that is still the concept of that singular thing, the intellect can conceive the singular through a more universal concept. If the object is of itself a singular, then the intellect would understand something through an account opposed to that very thing, as universality is opposed to singularity.

The nature of what is factically given has, on its own, a unity that is not the unity of singularity but less. Scotus calls this less-than-singularizing difference the common nature. Although the nature is not a singular, it is not, from that alone, universal. Its real unity is neither singularity nor universality.

> Understand: the nature is not of itself one by numeric unity; nor many by a plurality opposed to that unity; nor actually universal in the way in which something is made universal by the intellect, not as an understood object; nor is it particular of itself. For although it would never really be without one of these, nevertheless it is not of itself one of these, but is naturally prior to all of these.[32]

This common nature is precisely what is either universalized in understanding or singularized in the thing and, as such, it must be prior to and independent of either singularity or universality. Scotus acknowledges that the common nature "is never without" either singularity or universality, but that does not entail that it is either one or the other. That it is able to be universalized *or* singularized shows that, in itself, it is indifferent to each. We can say that, when it exists, it either exists in the soul, and then it is universal, or in the factically given, and then it is a singular. If, however, existence "contracts" it to either universality or singularity, then we must say that it does not exist independent of these, even though it is real. This argument, though Scotus does not make it explicit, shows that what counts as real must be more than what counts as existing, much less what counts as actual. That is, since the common nature has a real unity, it is real, yet Scotus argues that, when existing, it is either universal or singular. Therefore, the reality of the common nature stands outside its existence in the soul or in the thing. Furthermore, Scotus will call the existing singular "actual" but not the common nature. As a result, Scotus opens a notion of reality that extends beyond actuality and beyond existence.[33]

Scotus refers to this priority to universality and particularity as "indifference." According to him, when the common nature exists, that is, has *esse*, it has it either in the soul or in the particular thing. For Scotus, in fact, this distinction between *in anima* (in the soul) and *extra animam* is the first distinction of being.[34] Therefore, the common nature can be either universal*ized* or singular*ized*, but in itself it is indifferent to either of these. Scotus goes even one step further: it is not universalized simply by having *esse* in the intellect. Universality is one mode under which the common nature can be understood, but it is understood first of all in this indifference. What is more, "Universality is not part of its first concept, because universality is not a concept of metaphysics but of logic."[35] Therefore, Scotus insists that in relation to the common nature, a cause of *both* singularity and universality must be sought, as the nature is itself neither universal nor singular.

Yet equally difficult is the cause of singularity. Scotus argues that this common nature should be understood as: (1) the what-it-was of the thing whose nature it is; (2) the *per se* object of the intellect; (3) investigated by metaphysics; (3) expressed through the definition; (4) true, *per se* propositions are true on account of this quiddity. The quiddity, as common, would be a nature that belongs or can belong to many individuals. Thus, its commonality would be analogous to the commonality of the concept of being. However, we saw that the concept of being was "determinable," requiring another concept to determine it. Scotus addresses the difference that "contracts" the common nature to singularity by first noting that "just as unity in common follows *per se* some reality[36] in common, so too any unity follows per se some reality. Therefore, unity, such as is the unity of the individual . . . follows *per se* some reality."[37]

The argument follows from his position on the common nature, namely that if there is a unity that is not the unity of singularity, then that unity must be based on something, for it is a real, and not a merely rational, unity. Scotus pushes this same principle one step further: there must also be a reality from which the unity of singularity follows, otherwise singularity will be merely a construct of reason.

The factical is itself a unity having the nature. The reality from which singularity follows must account for two things at the same time: (1) it must determine the nature and (2) with the nature it must make something that is one. This is because, Scotus adds, "the whole whose unity this is, is perfect of itself."[38] This is the same structure we saw with respect to the concept of being and its relation to the concept of the ultimate difference. As being was merely determinable, so too is the common nature with regard to singularity. As the specific difference was merely determining, so too is the singularizing difference with regard to the nature.

As Scotus indicated with regard to the commonness of the concept of being, differentiation does not belong, essentially, to being, but comes, as it were, from another order. Being belongs to the determinable while difference is a determination of what is determinable. Yet the other side of this, a side that Scotus did not draw out, is that being as determinable must, as it were, be susceptible of differentiation. Anything that has been determined will be a being and, therefore, being can be predicated of it *in quid*. Since being is common to all things, it cannot itself be that determination. The very commonality of being precludes that it also is predicated of that which determines it. "Rational" is a determination of being, such that any factically given thing that is rational will be a being. Differentiation is "outside" of being and yet a differentiation *of* it. Even though being is nothing if it is not differentiated, it is not, of itself, productive of its own determinations.

Difference, therefore, is not conceivable in terms of anything other than itself. It is, for Scotus, irreducible or, to use his vocabulary, irresolvable. This irreducibility, in fact, is one of his main arguments that there must be a reality from which singularity follows. He begins by noting that there is always an immediate or primary diversity underlying the difference between things that differ.[39] If there were not some immediate diversity, "there would be no end to differences."[40] Differentiation would never end if there were not, ultimately, a diversity that is immediate, that is, simply diverse without further account. The nature, therefore, cannot be that from which the difference between individuals follows. Since the nature cannot account for the difference between individuals, there must be an additional reality belonging to the individuals on account of which they differ or are diverse. Scotus concludes, once again, by returning to the issue of determination: "therefore, there will some positive realities [*entitates*] determining, per se, the nature."[41]

The relation of determinable/determining is shown by comparing what is thus differentiated to what is below it, what is above it, and what is on the same level as it. In the case of the ultimate or specific difference, the specific difference has, in a sense, nothing below it, for "it is repugnant to the specific difference and to the specific reality per se to be divided, essentially, into many species or natures, and through this it is repugnant to the whole of which that specific difference is a *per se* part."[42] Each kind of unity is inconsistent with a division of some sort and to discover what kind of unity is involved, we need to discover what kind of division is repugnant to that unity. In the case of specific unity, it is inconsistent for it to be divided into many natures or species. Analogously, "to individual reality [*entitas*], it is repugnant to be immediately divided into any subjective parts,[43] and through this such a division is repugnant to the whole of which [the individual reality] is a part."[44] In the case of each difference, we see that, from the reality [*entitas*] from which the difference follows, there follows also a unity such that the whole of which the reality is a part resists division. The difference thus also constitutes a whole and because the difference is a reality belonging to that whole, it makes that whole resist division of one kind or another. So as regards what is below it, the individual difference constitutes, with the nature, a whole such that that whole resists being divided into parts.

In regard to what is "above" the difference, Scotus argues that the reality[45] from which the specific difference follows "is actual with respect to that reality from which the genus, or the account of the genus, is taken, such that this reality is not formally that, otherwise there would be idle chatter [*nugatio*] in the definition, and the genus alone would define sufficiently, because it would indicate the whole defined without the difference."[46] Here Scotus makes good on the determinable/potential and determining/actual relation. The specific difference brings, in its work of differentiation, a determination to the account of the genus and in this way brings an actuality to the potency of the genus. For Scotus, the general principle is that differentiation is actualization with regard to that which it differentiates. This also includes the difference that makes an individual, though his language becomes more tentative at this point: "[T]his reality of the individual is similar to the specific reality, because it is an as-if act determining that as-if possible and potential reality of the species, but as to this it is dissimilar, because the [individuating reality] is never taken from an added form, but is precisely the ultimate reality of the form."[47]

It is not easy to understand the repetition of the *quasi*, the "as-if" here. Scotus did not qualify to this extent his argument that the determinable is potential with regard to the actuality of the determining. The reality of the common nature constitutes the thing whose nature it is in what Scotus calls *quidditative esse*, the kind of existence that a nature or quiddity has. That kind of existence has been already established independently of factical singulars. Therefore, it must be kept apart from the understanding of the reality, existence, and actuality of any singular. In understanding the reality of the nature, we do not understand that from which any factically given thing would be "this." This means not only that the reality of the nature differs from the reality that makes the singular but also that they have different modes of existing, of *esse*. As I noted with regard to the concept of being and its determination, here too we must mark that the nature is indeed real but is not existing, except in *quidditative esse*, because in order to be actual is must be "contracted." As he interprets Aristotle, the nature, the quiddity, is the form and matter is that which has form. If we take a nature or quiddity, for example, humanity, we must note that it is not the form of one part or the other, of matter or of form, but "of the whole composite having a contracted quiddity. Therefore, every specific reality [*realitas*] constitutes in formal *esse*, because in quidditative *esse*, and the reality [*realitas*] of the individual constitutes precisely in material *esse*, that is, in contracted *esse*."[48]

This linking of material *esse* with the individuating reality, however, must be understood quite precisely. Scotus has already argued that matter cannot be the individuating reality, precisely because it needs its own principle of individuation. That is, if each human is made of sinew and bones, then that matter cannot be that by which any material individual is the individual it is. Furthermore, here Scotus has already insisted that the form, understood as quiddity, is the form of the whole that is a composite of matter and form. Since it is the form of matter, matter cannot be that from which singularity arises. Rather, Scotus is arguing that what is thought as matter is the contracting of the nature to singularity. Or, more precisely still, since the common nature has already been shown to be the object of understanding (though not first and adequate), Scotus is here using the term "matter" to mark that which is other than thought. There is a way in which matter, understood traditionally, is entirely recuperable by thought, particularly when it is posited as the principle of individuation—as it is in Aquinas. Scotus's argument here is that there is necessarily a reality that cannot be the object of thought that is the reality from which singularity follows. This reality that is other than thought is what Scotus calls its *esse material*.

In this way, Scotus makes a gesture toward a rethinking of matter away from understanding it as pure potency—a designation of matter that arises already in Aristotle. Here what he calls "matter" and material *esse* is that reality that contracts the nature to singularity. Yet, as I have argued, that contracting element is not potency, but is the ultimate actuality of the existing singular. While the common nature is certainly the quiddity of the thing, and therefore its essence, it is, for him, no "higher level essentiality," but rather a potency with regard to the actuality of the singularizing difference. The essence, the common nature, is not "extracted" from the thing. It is real only as determinable by that which is necessarily outside it and, therefore, outside of thought. Rather than the identity thought through the concept of the quiddity, it is difference, nonidentity, that brings existence and actuality to the factical. It is that very facticity, in the form of the individuating difference, that is the matter of thought, a matter that will always necessarily remain unthought. If the specific difference is a simply simple concept because there is no other concept into which it can be resolved, then the individuating reality will have no concept at all and will be simply that which is other than thought.

THE REDEMPTION OF SINGULARITY IN THE CONCEPT

As I argued in chapter 2, the contradiction that underlies and motivates Adorno's *Negative Dialectics* is that philosophy in its identitarian form cannot attain the very thing that set it in motion: thought should be of *something* and yet, in identifying that something with its thinkability, it pushes out of reach that something. The difficulty of this contradiction lies in the fact that if, on the one hand, I resolve it by reducing the reality of what is to that in it that is thinkable, I prevent thought from ever being about anything other than itself; but if, on the other hand, I resolve it by insisting on an immediate access to the factical, I also prevent thought from reflecting on the factical, I succumb to the "replacement of knowledge by the mere registering, ordering and summarizing of facts."[49]

This would be nothing more than a methodological problem for philosophy—and quite a serious one, at that—if it were not the case that "[o]ur metaphysical faculty is paralyzed because actual events have shattered the basis on which speculative metaphysical thought could be reconciled with experience."[50] Adorno's argument, at the end of *Negative Dialectics*, is that actual events that he names Auschwitz have both called a lie to and raised the need for metaphysical reflection.[51] The problem that is not merely methodological is that metaphysics has always been the implicit claim that existence, the factical, has meaning, a meaning that is only discovered in a beyond or in transcendence. Yet "actual events" make a mockery of that insistence upon meaning. However, on the other hand, to say to the victim that "this is all there is" would be an act of barbarism. There is, therefore, a promise of metaphysics that, on Adorno's argument, can only be achieved through a self-reflection of thought.

This self-reflection is the dialectics that Adorno calls "negative." Yet, it is also a metaphysics—he does not shy away from maintaining that term—that has materialism forced upon it.[52] However, the materialism that stands at the core of what Adorno names "a new categorical imperative" is not the claim that whatever is happens to be matter, nor the old metaphysical notion that matter is potency that is actualized by that element that responds to thought, the essence. Rather, it is a notion that there is a reality that is not identical to thought and about which thought attempts to turn. The actual events that have called traditional metaphysics into question demand that thought does not turn away from the factical. The experience of suffering, of a fate worse than death, demands a thinking of the factical. However, there is a thought of the factical, a nominalist/positivist approach to the factical, that affirms the absolute fungibility of every individual. For Adorno, materialism, the new metaphysics informed by the new categorical imperative, is just this thinking of facticity that does not, at the same time, undergird its absolute exchangeability. I will return to this in chapter 5; for this thought requires us to see how materialism necessarily implies a thought of the social conditions of its possibility.

NOTES

1. Theodor Adorno, *Metaphysics: Concept and Problems*, ed. Rolf Tiedmann, trans. Edmund Jephcott (Stanford: Stanford University Press, 2001), 18; Theodor Adorno, *Metaphysik: Begriff Und Probleme*, vol. 1, Nachgelassene Schriften: Vorlesungen (Frankfurt: Suhrkamp, 1998), 1:33.

2. Adorno, *Metaphysics: Concept and Problems*, 4; Adorno, *Metaphysik: Begriff Und Probleme*, 1:14.

3. Adorno, *Metaphysics: Concept and Problems*, 19; Adorno, *Metaphysik: Begriff Und Probleme*, 1:34.

4. Lawrence Dewan, "Obiectum: Notes on the Invention of a Word," *Archives d'Histoire Doctrinale et Littéraire de Moyen Age* 56 (1981): 37–96. According to his argument, *obiectum* is used to translate Aristotle's ἀντιχείμενα. In *De Anima*, Aristotle uses the term to name that to which a sense, for example a vision, responds. Color would be an example of an "object" where vision is concerned. Yet in this case, an object is not something "out there," independent of the soul, but that which arises when a capacity of the soul reaches out to the world and is met. For a slightly different take on this term and its translation, see Pietro Giuffrida, "Obiecta, Opposita, Antikeimena nel De Anima di Aristotele," *Giornale Di Metafisica* 2 (2013).

5. Ioannes Duns Scotus, *Quaestiones Super Libros Metaphysicorum Aristotelis Libri VI–IX*, ed. R. Andrews, G. Etzkorn, G. Gál, R. Green, F. Kelley, G. Marcil, T. Noone, and R. Wood, vol. IV, Opera Philosophica (St. Bonaventure: The Franciscan Institute, 1997), IV:515.

6. It is often difficult to render *esse* in English. While it is grammatically the infinitive "to be," it is frequently used in the Middle Ages as a substantive. "The to be," is not only awkward, it also might give a misleading impression. Translators frequently choose either "being" or "existence." Neither of these are good alternatives, as Scotus in particular has different notions of being and existence that are not the same as *esse*. For these reasons, I have left it untranslated.

7. Scotus, *Quaestiones Super Libros Metaphysicorum Aristotelis Libri VI–IX*, IV:518.

8. Ioannis Duns Scotus, *Ordinatio Liber Primus d.26–48*, vol. VI, Opera (Civitas Vaticana: Typis Polyglottis Vaticanis, 1963), VI:254.

9. Ibid., VI:296.

10. Ibid., VI:296–97.

11. Ioannes Duns Scotus, Balić, Carolus, Perantoni, Pacifico Maria, *Ordinatio liber primus, distinctio tertia*, (Civitas Vaticana: Typis Polyglottis Vaticanis, 1954), pars 1, q. II.

12. Ibid.

13. Ibid.

14. Ibid., q.3.

15. A "simply simple" concept is one that is not further resolvable into many concepts. Scotus gives two examples, the concept of being and the concept of an ultimate difference, that is, a specific difference.

16. Ibid.

17. Here I refer to Scotus's term *viator*, which is perhaps more literally translated as "wayfarer." It refers to the human after the fall and living without the beatific vision and living before the final resurrection.

18. Duns Scotus, *Ordinatio liber primus, distinctio tertia*, pars I, q. II.

19. Ibid.

20. Theodor Adorno, *Negative Dialektik: Jargon Der Eigentlichkeit,* vol. 6, Gesammelte Schriften (Frankfurt am Main: Suhrkamp, 1970), 6:44; Theodor Adorno, *Negative Dialectics,* trans. E. B Ashton (New York: Continuum, 1973), 34 (translation modified).

21. Duns Scotus, *Ordinatio liber primus, distinctio tertia,* pars I, q. II.

22. Ibid.

23. Ibid.

24. Ibid.

25. In *Difference and Repetition,* Deleuze claims "There has only been one ontological proposition: Being is univocal. There has only ever been one ontology, that of Duns Scotus, which gave being a single voice," Gilles Deleuze, *Différence et Repetition* (Paris: Presses Universitaires de France, 1968), 52; Gilles Deleuze, *Difference and Repetition* (New York: Columbia University Press, 1984), 35. Deleuze goes on to insist that "[Being] is said, in a single and same sense, *of* all its individuating difference and modalities," Ibid. 53/36. We can already see that this is not the case, that Scotus holds out a special role for differentiation in his ontology. In the following section, I argue that the individuating difference is, for Scotus, an even more extreme form of differencing, of which being is not predicated *in quid.* To my mind, it might just turn out that Deleuze is right for the wrong reason. He moves too quickly on the question of univocity only to conclude that differentiating is just what being is. Scotus's position, as we shall see, might amount to just the same thing.

26. It is also not predicated *in quid* of the attributes of being such as "one" or "true." I will leave aside the discussion of the attributes of being.

27. Scotus, op. cit.

28. Duns Scotus, *Ordinatio liber primus, distinctio tertia,* pars I, q. 2.

29. Ibid.

30. Ibid.

31. Ioannis Duns Scotus, *Iohanni Duns Scoti Opera Omnia,* vol. VII, n. 7 (Civita: Typis Polyglottis Vaticanis, 1973); Peter King, "Duns Scotus: Ordinatio II D. 3 p.1 q.1," 1987, http://individual.utoronto.ca/pking/translations/SCOTUS.Ord2d3p2q1.trns.pdf, p. 7.

32. Scotus, *Iohanni Duns Scoti Opera Omnia,* vol. VII, n. 7; King, "Duns Scotus: Ordinatio II D. 3 p.1 q.1," 7.

33. This, it seems, is the appeal that Scotus held for C. S. Peirce. While Peirce eventually gave up even on Scotistic realism as being too close to nominalism, his initial receptivity to it seems to be intimately connected to his distinction between firstness, secondness, and thirdness, that is, a distinction that itself maintains a notion of the real beyond the actual and beyond the existent. On this issue, see John F. Boler, *Charles Peirce and Scholastic Realism, a Study of Peirce's Relation to John Duns Scotus* (Seattle: University of Washington Press, 1963).

34. Steven Marrone, "Henry of Ghent and Duns Scotus on the Knowledge of Being," *Speculum* 63, no. 1 (1988): 22–57.

35. Scotus, op. cit.

36. The term Scotus uses is *entitas,* which King translates literally as "beingness." I do not quibble with this translation, as it is probably less interpretive than my choice of "reality." However, I choose this translation for several reasons. First, it is clear that in this context Scotus is pointing to that which does not necessarily exist, and yet it has a reality of some sort that is more than the reality of a being of reason. Second, in his discussion of the Trinity, Scotus uses the term frequently in order to describe how it is that Father, Son, and Holy Spirit are in actuality one and the same, even though there is a difference to be made.

As he says in the course of that discussion, "The first supposit formally or really has a communicable *entitas* . . . And I understand 'really' is this way, that in no way through an act of the considering intellect, or rather that such *entitas* would be there if no intellect were to consider it," Scotus, *Ordinatio, Liber Primus, D. 1–2*, vol. II, n. 390. Here Scotus clearly links the notion of the real with that of *entitas* and precisely to indicate a kind of reality that is not belonging to the intellect alone but also is not found as a "thing."

37. Scotus, *Iohanni Duns Scoti Opera Omnia*, vol. VII, n. 169; King, "Duns Scotus: Ordinatio 2 D. 3 P. 1 Qq. 5–6," 9.

38. Scotus, *Iohanni Duns Scoti Opera Omnia*, vol. VII, n. 169; King, "Duns Scotus: Ordinatio 2 D. 3 P. 1 Qq. 5–6," 10.

39. An immediate or primary diversity would be between two things that are simply different in such a way that the difference cannot be resolved into something that could be common to both or even to some third thing.

40. Scotus, *Iohanni Duns Scoti Opera Omnia*, vol. VII, n. 170; King, "Duns Scotus: Ordinatio 2 D. 3 P. 1 Qq. 5–6," 10.

41. Scotus, *Iohanni Duns Scoti Opera Omnia*, vol. VII, n. 170; King, "Duns Scotus: Ordinatio 2 D. 3 P. 1 Qq. 5–6," 10.

42. Scotus, *Iohanni Duns Scoti Opera Omnia*, vol. VII, n. 177; King, "Duns Scotus: Ordinatio 2 D. 3 P. 1 Qq. 5–6," 12.

43. It should be noted that in medieval terms, "subjective" indicates the kind of being that a *subiectum*, that is, ὑποκείμενον, has. That is, it is almost opposite of the modern (at least since Kant) notion of "subjective." This is the other side of what was indicated above in note 6, see Dewan, "Obiectum: Notes on the Invention of a Word."

44. Scotus, *Iohanni Duns Scoti Opera Omnia*, vol. VII, n. 177; King, "Duns Scotus: Ordinatio 2 D. 3 P. 1 Qq. 5–6," 12.

45. Here Scotus uses the term *realitas* in the exact place he had used *entitas* previously.

46. Scotus, *Iohanni Duns Scoti Opera Omnia*, vol. VII, n. 179; King, "Duns Scotus: Ordinatio 2 D. 3 P. 1 Qq. 5–6," 12.

47. Scotus, *Iohanni Duns Scoti Opera Omnia*, vol. VII, n. 180; King, "Duns Scotus: Ordinatio 2 D. 3 P. 1 Qq. 5–6," 13.

48. Scotus, *Iohanni Duns Scoti Opera Omnia*, vol. VII, n. 182; King, "Duns Scotus: Ordinatio 2 D. 3 P. 1 Qq. 5–6," 13.

49. Adorno, *Metaphysik: Begriff Und Probleme*, 1:179; Adorno, *Metaphysics: Concept and Problems*, 115.

50. Adorno, *Negative Dialektik*, 6:354; Adorno, *Negative Dialectics*, 362.

51. In the lecture course, *Metaphysics: Concept and Problems*, Adorno clarifies what he is thinking under this proper name: "Through Auschwitz—and by that I mean not only Auschwitz but the world of torture which has continued to exist after Auschwitz and of which we are receiving the most horrifying report from Vietnam—through all this the concept of metaphysics has been changed to its innermost core." Adorno, *Metaphysik: Begriff Und Probleme*, 1:160; Adorno, *Metaphysics: Concept and Problems*, 101. This course took place in 1965.

52. Adorno, *Negative Dialektik*, 6:358; Adorno, *Negative Dialectics*, 365.

FOUR

Hobbes on Materialism as Social Philosophy

If materialism is not somehow involved with concepts and with reason, then it turns out to be what is normally and more properly called myth or theology. Aristotle saw this clearly. If the investigation of nature does not appeal to thought, then "everything is from Night."[1] There is, then, a tremendous difference between Hesiod's assertion "First came Chaos," and Thales's position "All is from water," that is, a difference between a mythical positing of a being that, as such, stands beyond reason, and the investigation into the principles that reason can discover.

If materialism is to avoid this mythic positing, then the difference between it and its other(s)—rationalism or idealism, for example—is not as great as might be suspected. The difference, however, cannot be one of mere positing: idealism posits ideas as metaphysically basic while materialism rejects them. Rather, materialism, if it is not theologically naïve or mythic, must be the rational, conceptual insistence that we have no right to the assumption that what is ultimately real must correspond to our concepts. If the thesis of this correspondence, which Adorno calls "identitarian thinking," is what constitutes metaphysics as such, then materialism turns out not to be a metaphysics at all, but rather a non- or at least parametaphysics. As I indicated in chapter 3, Adorno identifies Aristotle as the first metaphysician for two main reasons. First, Aristotle turns what corresponds to the concept—form or essence—into the "higher level essentiality"[2] of phenomena. This position, however, he shares to a large extent with Plato. Thus, the second feature of Aristotelian thought that makes him the first metaphysician is that he, unlike Plato, focuses on the mediation between phenomenal reality and its higher-level essentiality. Unlike Heidegger, Adorno does not simply critique metaphysics. Rather, for Adorno metaphysics is always, at least partially, critical in itself. There is something also to be rescued in metaphysics. The drive to think the factical is precisely what is deserving of rescue in metaphysics. The critique of metaphysics as positing an identity—indeed of an "appropriation without equivalence"—must not turn away from the thought of the factical. If this is what constitutes metaphysics, then a materialism that is not mythic or theological will stand outside metaphysics. And yet, it cannot simply be posited as standing outside metaphysics. It must emerge, rather, from out of a critique of metaphysics, that is, from a resistance to this identitarian thinking.

The thinking of materialism in Scotus, Marx, Adorno, and Althusser might go some way toward explaining what might appear as an anomaly in Hobbes's thought.[3] One the one hand, Hobbes certainly does maintain whatever is, is a body and every body is matter. On the other hand, the crucial term involved here, "body," is secured by what Hobbes calls "the work of reason," and, as such, is never reduced by him to material principles. While this purported anomaly is often used as a critique of Hobbes, showing a fundamental disjunction between his normative philosophy[4] (ethics and politics) and his mechanistic materialism, it might actually show the reverse—the fundamentally normative nature of Hobbes's materialism and, therefore, the fundamental unity of his philosophy. The proper sense of materialism in Hobbes's philosophy can be exposed by looking at how he critiques Aristotelian metaphysics and the response he presents to this critique. This presents an itinerary that ought to be followed in order to expose just how Hobbes is a materialist. The road begins with Hobbes's own statement that whatever is, is a body. What are the conditions for the possibility, within his thought, for such a statement? This will require that I travel through what Hobbes calls *prima philosophia*—an obvious recasting of Aristotelian metaphysics outside of a commitment to substantial forms. However, I will show that the achievements of first philosophy are conceptual or rational, and thus require an investigation into Hobbes's understanding of reason. Finally, the road can double back to the critical moment in Hobbes's political philosophy—the covenant—to show that the use of reason there is in fact a result of his materialism—a materialism that does not shy away from the "work of reason" in thinking the factical.

WHATEVER IS, IS A BODY

If Hobbes is a materialist, so a story goes, it is because he maintains that being is matter. For Hobbes, being (*ens*) is "that which occupies space, or that which can be measured by length, width, and depth. From which definition it appears that being and body are the same; for the same definition of body is indeed received by all."[5] Hobbes then goes on to equate body and matter: "Body, however, and matter are names of the same thing nevertheless on account of diverse considerations, indeed the same existing thing considered simply is called body, considered as susceptible of a new form or new figure is called matter."[6] By means of a simple syllogism, we can conclude that, for Hobbes, what is, is matter. Because of this simple syllogism, it is obvious why Hobbes's first philosophy does not treat the term "being" at all. More specifically, it is clear that when Hobbes comes to treat "body," he is, in fact, treating being as such.

Hobbes's position, however, is complicated by the fact that he is only able to define being as body and matter by excluding any consideration of beings that "exceed the grasp of humans."[7] The exclusion of such beings, "which are normally called incorporeal substances," is based on the idea that "it is permitted for the philosopher least of all to posit and dispute about those things that exceed the grasp of humans."[8] So the equation of being with body and with matter arises not directly from any argument about the proper meaning of the term "being" or about what we can know about any being, but from a critique of the philosophy that would posit nonbodily substance. For this reason, before Hobbes equates being with "body," he has already equated it with "imaginable." Yet these two terms are not identical in the way that being and body are. Philosophy deals with what is imaginable, that is, within the grasp of humans, and it turns out that the only imaginable kind of being is a body and hence material.

But what is the source of this prohibition? If one assumes that the restriction of being to the imaginable results from Hobbes's theory of knowledge—a theory that begins with sensation and imagination—then Hobbes is caught in a circle. Being is body because only bodies can be sensed (and therefore imagined) and only bodies can be sensed because otherwise there would be dispute about noncorporeal beings. Less circularly, but still equally vicious, one could assume that the metaphysical posit that being is matter rests on an epistemological posit that all knowledge begins in sensation and only bodies can be sensed. Yet if the position that whatever is, is matter rests on such posits, then Hobbes's materialism is not very different from mythology ("First came Chaos.").

This is why "it is permitted for the philosopher least of all" to treat matter as if it were a god, even a chthonic one. Here Hobbes implicitly recalls Aristotle's errant path—the very errancy of matter as such, of materiality. If, on the one hand, matter is posited as simply what is, then matter is a principle like Night. In fact, Aristotle had already equated the mythological notion that everything is from Night with the materialists' (beginning with Thales) claim that all things are from a material origin and all things were initially together. The difficulty is that, without a cause, one cannot account for how everything was set in motion. "Matter is not going to set itself in motion—its movement depends on a motive cause, such as carpentry."[9] Yet, on the other hand, when the philosopher comes to posit reason—that which is thinkable in what is—as the reality or higher level essentiality of what is, the philosopher turns that into another goddess from which everything comes. Hobbes's insistence that the philosopher, least of all, is permitted to posit as an explanation that which "exceeds the grasp of humans" must, therefore, work in both directions, that is, against matter and against substantial forms, as both would be equally cthonic deities.

Hobbes's restriction of being to the imaginable, that is, to what does not exceed our grasp, rests not on a metaphysical or epistemological posit, but on a social and political analysis of the role of philosophy in civil strife. The *Leviathan* ends with a discussion of what Hobbes calls the "kingdome of Darknesse." While on the face of it, this section is meant to show how a commonwealth can come to be ruled by Satan, "the Prince of the Power of the air,"[10] it also lays bare the seeds that bear the fruit of social discord. The kingdom of darkness is "a confederacy of Deceivers, that to obtain dominion over men in this present world, endeavour by dark, and erroneous Doctrines, to extinguish in them that light, both of Nature and of the Gospell; and so to dis-prepare them for the Kingdome of God to come" (Lev. 44, 215). At stake here is the move these deceivers make in order to establish dominion over people. The question, for Hobbes, is always *cui bono*? If we are turned away from the light of nature toward ghostlike realities—such as separated essences—then the first step in understanding how this happens is to understand who benefits from the reality of ghosts. These deceivers need to turn people away from reason—the light of Nature—toward darkness. Hobbes's concern, therefore, is not so much the validity or invalidity of particular doctrines, but the way in which theological and philosophical positions bring about darkness and, consequently, sow the seeds of civil war.

> But to what purpose (may some man say) is such subtilty in a work of this nature, where I pretend nothing but what is necessary to the doctrine of Government and Obedience? It is to this purpose, that men may no longer suffer themselves to be abused, by them, that by this doctrine of *Separated Essences*, built on the vain philosophy of Aristotle, would fright them from Obeying the Laws of their Countrey, with empty names; as men fright birds from the corn with an empty doublet, a hat, and a crooked stick. (Lev. 46, 238)

The critique of Aristotelian "vain philosophy" rests, therefore, on this concrete, sociopolitical analysis. This critique leads Hobbes to restrict philosophy to the imaginable, which in turn leads him to equate being with matter. In this way, the position that being is matter rests on a more fundamental materialism, a materialism that has sociopolitical structures as its subject.[11] This, therefore, might be the proper place to call Hobbes a materialist, for this is an analysis of the concrete social and political source of and effects of ideas. That is, Hobbes does not treat philosophical concepts as immediately related to a "higher level essentiality," but rather as generated for a certain social, economic, and political structure. The problem of Aristotelian philosophy that Hobbes diagnoses here is one of the very grounds of political discourse. The appeal to separated essences leads to a discussion of virtues that allows for subjects to lessen their dependence on the sovereign power of their country. "For who will endeavour to obey the Laws, if he expect Obedience to be Powered or Blown into him" (Lev. 46, 328). That is, if being good depends on an essence that is not governed by the laws that govern material bodies, then I cannot expect to be good unless that goodness possesses or haunts me as a spirit. The only legitimate ground of political discourse should be that which is equally available to all: reason and what appears to the senses. Such a ground would make political discourse the domain of all and would eliminate the benefit that those who propound Aristotelian essence receive from their deception. In this way, the metaphysical position "whatever is, is matter," expresses Hobbes's commitment not to a posited mechanistic materialism but to the sociopolitical materialism that traces the origin of ideas by asking the question "cui bono?"

We are, once again, on familiar terrain. Hobbes is attempting to come to an understanding of what is, of the factical, that fully accounts for the reality of ghosts. Yet, precisely because of the operation of spirits, we cannot refuse the analytic power of concepts. In order to think matter, one needs to have recourse to concepts that are not reducible to materiality; otherwise the claim that being is matter is methodologically identical to the claim that "Everything is from Night." That Hobbes is aware of this methodological difficulty is clear from the very structure of his *De Corpore.*[12] The fundamental claim "being is to be a body" is only grounded in Part II of the text. Prior to that is a discussion on logic that culminates in a discussion of method. That is, the structure of the text recognizes that there is, on the one hand, a primacy belonging to the notion of matter and body and, on the other, a methodological dependency—it is ontologically prior and yet methodologically a result.

BODY AND PRIMA PHILOSOPHIA

Hobbes's treatment of first philosophy is not the first part of his *De Corpore*. Rather, it follows the part that Hobbes calls "*computatio sive logica*." This prior part deals with the definition, scope, and end of philosophy, as well as with names, propositions, syllogisms, and method. Part II, although entitled "*philosophia prima*," begins by naming the subject matter of this part of philosophy "*doctrina naturalis*." In other words, first philosophy has as its task the uncovering of the fundamental concepts that are required in a philosophical analysis of nature. The title of Part IV, "*physica sive naturae phaenomena*," treats nature as appearing (phenomenal) and thus differs from the treatment of certain "*doctrina*" that pertain to nature. Thus, while Part II (first philosophy) and Part IV (physics or the phenomena of nature) both deal, in one way or another, with "nature," they do so in different ways. And yet there seems to be a strange semicircularity in these two approaches to the topic of nature. On the one hand, first philosophy requires, as I will show, the presence of images and memory, which, for Hobbes, are grounded in sensation, even though it subtracts from these.[13] On the other hand, the analysis of sensation in Part IV requires concepts that are uncovered in Part II. The structure of *De Corpore*, therefore, indicates a difference between what presents itself[14] to sensation and what is required in order to analyze it. This distinction prefigures, in many ways, the discussion in Marx of the difference between the "chaotic representation of the whole" and the real concrete.

Hobbes's first philosophy takes its start by drawing away from nature as it appears. "We best take the beginning of the study of nature [*doctrina naturalis*] . . . from privation, i.e., from the fictive withdrawal [*sublatio*] of the universe."[15] The supposed annihilation of the universe presupposes a universe of things, one that we then suppose to be removed, taken away, or suspended. The result of this supposed annihilation is that the things that seem so obviously there to sensation are removed, and yet we can still ask what remains concerning which someone (excepted from the annihilation) could rationcinate, philosophize, and impose names for the purpose of ratiocinating.[16] The fiction of supposing the universe withdrawn is, therefore, a methodological tool that allows one to stand outside the flow of sensation in order to find what is available only to the powers of reasoning.

This methodological move shares features with Descartes' hyperbolic doubt.[17] Descartes, like Hobbes and Aristotle, recognizes that existing things do not bear on their face, that is, on the aspect that presents itself to experience, their own account. It is because of this separation between the way in which existing things present themselves and their account of "what they are" that makes it possible for the two not to be identical. Things can always present themselves as other than they are: "I have admitted the false as the true."[18] The solution to this problem, according to Descartes, is not to search all that I hold to be true and discover whether it is false—a process that will be without end. Rather, the solution is to withhold assent whenever there is the slightest reason for doubt. The notorious "evil genius" provides a reason why everything I take as true might turn out to be false, that is, everything that is might not be as it seems. Therefore, I must withhold assent from everything that claims to be true, that is, that claims to be a statement of what is.

The consequence of this is that I am without a world at all. Indeed, by the close of the First Meditation and the opening of the Second, the meditator is not even able to accept that she is. This move allows Descartes to step outside of the metaphysical pretensions of experience. It is precisely the "intentional" fact of sensory experience, indeed of all thought, that presents the problem with which the *Meditations* opens. Since sensation, judgment, willing, and so on are always *of* something, it carries with it the implicit presumption that the "object" of this act is other than the act itself. That something appears provides no ground for the claim that it is as it appears or is at all. It is this latter move, however, that is not allowed by Hobbes's fictive annihilation of the universe. While Hobbes's fictive move certainly also puts a stop to the metaphysical pretensions of experience, it remains committed to the world. That is, the annihilation of the world is precisely a withdrawal of *the world*, and therefore is parasitic on the givenness of the world to sensory experience. Descartes' methodological doubt means that the meditator cannot be sure that there ever was a world at all. Hobbes's fictive annihilation leaves one with the recognized absence of a world that once was.

This, in fact, points to the single, major difference between Hobbes and Descartes. As will be obvious, Descartes and Hobbes both insist on the power of reason to uncover the mode of operation of sensation. In this sense, they are both rationalists as much as they are both empiricists. However, Hobbes's fictive annihilation of the world never allows the supposition that the world as such might not be at all. He takes off from the presumption of the world and our being-in-the-world. Yet, like Descartes, he recognizes that being-in-the-world means being affected constantly by sensations that carry within themselves a metaphysical pretension. The only way out of that pretension is to put a stop to the flow of sensation. The fictive annihilation of the world does just this but so as to understand *that very world*. Descartes' methodological doubt turns thought out of and away from the world into thought as such. Being left without a world, the *cogito* finds it increasingly difficult to think something other than itself.

This difference between the methodological moves can be seen in what seems to be a marginal insertion in Hobbes's Objections to the *Meditations*. "I am a thinking thing, right. For from the fact that I think, *or I have phantasms*, whether being awake or asleep, it is gathered that I am a thinking thing."[19] In his response, Descartes seems not to notice the significance of replacing "thinking" with "having phantasms." Descartes' methodological doubt leads to the conclusion that I am able to think without anything other than thought that is the object of thought. Hobbes's quick replacement of "thinking" with "having phantasms" means that thinking will always be of something that is other than thought. Since a phantasm is, for Hobbes "nothing but *decaying sense*" (Lev. II, 13), it, like sensation, is always tied to something that, through moving, is its cause. The phantasm is possible only if there is, or has been, a world. There is, therefore, always a world outside of thought whenever there is thought. Thinking as such requires a world. Hobbes concludes this particular objection by drawing out the irresolvable difference between his position and that of Descartes:

> Because, therefore, the knowledge [*notitia*] of this proposition, "I exist," depends on the knowledge of this, "I think," and the knowledge of this latter from the fact that we cannot separate thought from matter of thinking, it seems it should better be inferred that the thinking thing is material than immaterial.[20]

It is the fact that, according to Hobbes, thought is nothing but an image that—as I will show below—is always caused by the motion of some body that leads Hobbes to the conclusion that thinking is material. Since thought cannot be without an object—a material body—and yet the body can be independent of thought, the image or phantasm will always have this "other." This required other is what Hobbes calls the matter of thinking. However, Hobbes will always agree with Descartes that *the way* the world shows itself does not appear in that showing. This is the point at which we need to put a halt to the showing, even if fictive, in order to uncover the *way* of that showing.

For Hobbes, what remains after the supposed annihilation of the universe are ideas of the world and of all bodies that were seen by the eyes or perceived by the other senses. In short, before tracing the way in which sensation is the cause of ideas (a task that belongs to physics, not first philosophy), Hobbes investigates what must belong to these ideas as such—without the things to which they seem to refer immediately. While sensation includes already within it an automatic reference to some thing outside the sentient, the supposed annihilation of the universe cuts off this seemingly immediate relation to the thing. It is important to note that what remains are not ideas of this or that thing, but rather "the memory and imagination of magnitude, motion, sound, color, etc., and also their order and parts" (De Corp. II, 7, 75). Such "accidents" are what was sensed before the supposed annihilation, and, therefore, once the automatic reference to the things sensed is removed, what remains are only the accidents that were sensed, or, more precisely, the effects of those accidents.

Since the universe is supposed to be annihilated, these ideas and images must be accidents not of things, but internal to the one imagining. However, "they will show themselves as external and not at all depending on a power of the soul" (Ibid.). Such ideas and images in themselves, that is, supposing the things to which they refer to be annihilated, still carry within themselves this "as outside." The fiction provides a way to think what belongs to ideas or images as ideas or images. This includes the "as outside," while allowing Hobbes to consider them, as it were, within thought. These ideas are that for which names are imposed and about which one ratiocinates.[21] The work of reason, composing and dividing, can work only on these phantasms.[22] Yet this "as outside" must be understood as belonging to ideas *as ideas*, since the things outside are supposed to be annihilated. "If we compute the magnitudes and motions of the heavens and the earth, we do not ascend to the heavens, such that we divide it into parts or measure its motion, but we make it while at rest, in silence or in darkness" (De Corp. II, 7, 76). What emerges here, on the basis of the supposed annihilation of the world, is the possibility of a kind of pause in the flow of bodies that causes sensation in the sentient. When Hobbes comes to analyze both sensation and the appearance of nature, he will argue that rest is always only apparent, that is, if something appears, it must in one way or another be in motion. Yet here, the supposed annihilation of the universe makes possible a stop in that motion and in the flow of that motion. We must suppose, on the basis of this thought experiment, that there are no things, and therefore no things in motion that are causing these phantasms. Being at rest, in silence and darkness, we make computations. The supposition of the annihilation of the world, we might say, interrupts the natural motion of all things and, therefore, allows one to consider ideas outside their natural and automatic reference to their origin in the motion of bodies. Since that motion causes nature to appear, whatever is treated in first philosophy will be independent of the way of appearing, that is, the phenomenality, of nature. In Part IV of *De Corpore*, phantasms will be traced causally back to their (merely) possible origin in the motion of bodies, and thus will show themselves to be a kind of motion. It is precisely this motion of phantasms that is paused here, and what emerges is the consideration of phantasms not as internal accidents of the soul, but as "the appearance [*species*] of external things, i.e., as non-existing, but appearing to exist or to stand outside" (Ibid.).

This marks the difference between first philosophy and the analysis of nature that comes in Part IV of the *De Corpore*; for in Part IV, sensation is traced to its possible causal origin in motion. As such, sensation is part of physics, since physics or natural philosophy deals with the appearing of things. One might say that physics analyzes things as appearing, that is, as phenomenally given. That analysis, as it were, stays within that phenomenality and traces the lines of causation that show how things appear. Here, in first philosophy, Hobbes is attempting to find a way to account for that very phenomenality, or, more precisely, to uncover those key concepts that will allow the tracing of lines of causation that give things to us phenomenally. The possibility of providing an account of sensation emerges only when the phenomenality of this is cut off, stopped, brought to rest. If scholastic-Aristotelian metaphysics uncovers concepts that lie behind nature as it appears, it does so by means of assuming an identity between our concepts of things and the higher-level essentiality that gives those things to appear. Since Hobbes's sociopolitical critique prevents that identity, his first philosophy proceeds by an assumption of a pause in the flow of phenomenality. The concepts uncovered in this way (space, body, cause, effect, potency, actuality, etc.) do not require an identity with what is real behind appearance. Rather, such concepts allow one to construct an account, merely possible and not actual, of how things as such might appear.

For example, Hobbes defines space as "the phantasm of the existing thing as existing, that is, having considered no other accident of that thing, other than that it appears outside the imagining" (De Corp. II, 7, 76–77). The possibility of "considering" a thing without considering an accident of it is not given within sensation, for sensation is always *of* accidents. If sensation is always of accidents, then the possibility of "considering" a thing merely as existing, that is, of imagining a thing without those accidents that are sensed, cannot belong to or derive from sensation. There is, in short, a capacity that operates upon the ideas that are materially caused by bodies, but is not, for that very reason, also materially caused. Ideas are both material things—insofar as they are accidents of the body of the sentient—and caused by the motion of bodies. The consideration of the phantasm merely as existing, however, cannot be a material thing nor caused by the motion of a body.

Below, I will unpack Hobbes's analysis of sensation in which he argues that sensation is an effect of the motion of bodies. However, as an effect, it should be produced when its causes are present. In fact, he insists that "the cause of every effect consists in certain accidents of the agents and patients, which, when all would be present, the effect is produced and if any are absent [*desit*], it is not produced" (De Corp. II, 9, 3, 96). Since the effect just is the sensation of accidents and accidents are "[faculties of bodies] by which they impress their concepts on us" (De Corp. II, 8,2,83), the "consideration" of body without considering its accidents cannot belong to sensation—a further capacity of activity must be involved. Furthermore, sensation is, as will be made clear, an effect of bodies and, therefore, material. This further capacity or activity, however, cannot be reduced to matter without running into an infinite regress or vicious circle.

This capacity for considering bodies outside the causal flow of phenomenality, however, does not presuppose the existence of something that does not appear, and thus something immaterial. Rather, it presupposes only the interruption of that flow so that one may consider the phantasm of a thing in a way other than as it appears in sensation. Therefore, a concept such as space and the subsequent concepts of first philosophy—body, accident, cause, effect, and so on—emerge as belonging to our phantasms as phantasms without reference to the causal origin in the motion of bodies. These concepts are required for Hobbes to then go on (in Part IV) to trace and analyze the causal flow of phenomenality. Their priority, their firstness, is therefore not a temporal one in our sensory experience—we do not first sense "body" and then this chair—but rather an analytical one in that we cannot give an account of the sensation of bodies without them.

"Whatever is, is a body" includes a term that belongs to Hobbesian first philosophy, namely "body." That is, it is a statement made not about the sensation of a thing but about how I can come to analyze that sensation. I never sense body even while everything that appears to sensation is a body. "Body," as I have shown, is not an abstraction from sensed things, but a *re*-traction—a removal of sensed things by means of the supposed annihilation of the world. Unlike space, which emerges immediately after the supposed annihilation of the universe, "body" requires an additional supposition. "We next suppose something of those [things supposed to be annihilated] to be placed back or created anew. It is therefore necessary that that created or replaced thing would not only occupy some part of said space, or be coincident and coextensive with it, but also to be something that does not depend on our imagination. This, however, is the very thing that is usually termed *body* on account of a certain extension" (De Corp. II, 8, 82).

Here the pause in the flow of appearance is even more clear than in the emergence of the concept of space; for body emerges *as* the independence of a thing from our images of it—an independence that never appears as such. This independence results from considering what would have to be the case if, after supposing the annihilation of the universe, we now suppose the placement or creation of anything whatsoever. Retracting the causal lines that bring this into being and retracting the causal lines that give this to appear to me, I am left with nothing but its standing there in space not depending on my having an idea of it. The idea "body," therefore, is not at all the same as a sensed body, nor even caused by the sensation of a body, since the idea of body is nothing more than the independence of a thing from my sensation.

As a result, the claim "whatever is, is a body" cannot be a claim about sensation or what is sensed. What is more, it is also not a universal name, as Hobbes defines universal names in Part I of the *De Corpore*. For Hobbes, a common or universal name is still the name of singular things "taken one by one, not, however, collectively all at the same time" (De Corp. I, 2, 9, 23). Hobbes's argument is directed against a universal term like "human" being the name of a genus or a species, that is, naming something like an essence that is the reality of the material thing. If "human" would be the name of the species, then, it would seem, there is something other than individual humans, some existing thing, of which "human" would be a name.[23] Hobbes goes even further: such universal terms are not names of ideas or phantasms, existing in the soul, of something, but are always names of spoken words or names (De Corp. I, 2, 9, 23). As names of singular things, they have "corresponding to them in the soul concepts, ideas, and phantasms . . . of singular things" (De Corp. I, 2, 9, 24). In the end, the universal has no reality other than that of the individuals that are taken, one by one, in the universal name. The only concepts corresponding to such names are concepts of singulars.

The concept of space and, consequently, the concept of body, however, are both general in the sense that they are concepts of more than one thing. Yet, they are not universals in this sense because they are not signs of existing, singular things. Indeed, since common or universal names are signs, like every sign they do not signify things but rather they signify that the one offering the sign had a thought (*cogitatio*) (De Corp. I, 2, 4, 21). Signs are always signs of concepts. As has been shown, however, concepts are "impressed in us" by those faculties of bodies Hobbes calls accidents. The definition of space as "the phantasm of an existing thing . . . having considered no other accident of the thing" defines, therefore, the consideration of something only as outside. Since we have this concept, it is impressed on us by some capacity or power [*facultas*]. Yet the only power of the thing considered is its being outside. So there must be another route to concepts other than their being impressed on us by the power of a body. That route comes through the fictive annihilation, that is, from the methodological withdrawal of the word. The term "body" uncovers, by means of the dual supposition of annihilation and restoration, what must belong to a thing place back in a world supposed to be annihilated.

This explains the difference in the definition of space, on the one hand, and body, on the other. Space is defined as the "*phantasm* of an existing thing insofar as existing, that is, having considered no other accident of the thing, other than that it appears outside imagining" (De Corp. II, 7, 76–77; emphasis mine). Body, however, is not defined as a phantasm at all, but as anything that, not depending on our cognition, takes up some part of space. In this way, body is a "whatever" [*quicquid*] that emerges as independent once the phantasm of space is supposed. Since body is the most fundamental "whatever" of first philosophy, it turns out that first philosophy is not about things as they appear, but about how we might conceive their mode of appearing.[24] The analysis of the mode of appearing cannot be collapsed into the analysis of appearing, for then Hobbes's "whatever is, is matter" would be formally identical to the Aristotelian separated substances and essences.

The supposition of the annihilation of the world, however, does not seem to stand on materialist principles. This supposition is, according to Yves-Charles Zarka, a hyperbolic assumption that, unlike Descartes' hyperbolic doubt, consists in "a passage from real to possible and not from doubtful to false."[25] Zarka goes on to point out that the supposition of the annihilation of the world implicitly reproduces William of Ockham's arguments concerning God's omnipotence. For Ockham, like Duns Scotus, the absolute power of God cannot extend to any act that would bring about a contradiction. For others, like Pierre d'Ailly, this power is prior in a certain way even to the principle of noncontradiction.[26] The distinction between these two ways of understanding God's power—either as absolute or as acting through the order God has instituted—effects a certain possibility of standing outside the natural order that God has, in fact, instituted. If I ask, for example, whether my knowledge is always of an existing and present object, I must answer that given the conditions God has chosen to create, this must always be the case. However, this answer does not tell me what knowledge *as such* is. In order to take up that question, I must be able to suspend, at least suppositionally, the order that God has, in fact, created and consider the potentially infinite possibilities open to the divine will. This is brought about by considering God's absolute power.

This description, however, presents the methodological use to which this position can be put, a use that is also operative in Hobbes's hypothesis of an annihilated universe. The distinction itself, however, does not arise initially in medieval thought as a methodological tool, but as a way to understand the divine attribute of omnipotence. If we understand God's creative activity on the basis of the order God has established, we impose limits on that very creative activity. If, for example, we take the Aristotelian position that nothing can be without form, we assume, therefore, that creation was merely the choice of *which* forms to bring into being. Indeed, implicit in Ockham's insistence on the absolute power of God is a critique of the correlation between the being of things (form) and our knowledge of them (the form in the soul). Such a view of creation, however, is contrary to the concept of divine omnipotence. If God is truly omnipotent, then God must be understood to create out of God's will alone, without any prior condition. The absolute power of God is the power to bring something to be by means of a will that is determined by nothing other than its willing. The determination of things—either in knowledge or being—is consequent on the divine will, but that will itself is not determined by anything prior.

This voluntarist tradition certainly has political implications, particularly when late medieval thinkers consider pope and king as analogous to God in the relation between laws and conditions prior to any law.[27] It also has implications for theories of knowing and being, and for the task of philosophy as a whole. Positing a being who creates the universe from an act of its will—a will determined only by itself and not by reason, justice, or the good—leaves us, as it were, on this side of phenomena without the possibility of arguing that phenomena are the mere appearance of a higher-level essentiality that does not appear. In other words, since God is the ultimate cause of all things and yet causes only by willing, we can only point to that will as cause and not some further "reason" why God causes this world and not any other. The insistence that God is an intelligent, that is, rational, cause and that the essence of each thing is the "reason" or account of its being at all turns the divine will into a mere effective instrument of the divine intellect, which differs from ours not in terms of object but only in terms of scope. However, even granting God's absolute power, one may still attend to the structures that determine this world that God has created as well as to the general conditions that would have to belong to any creation at all. Even granted the absolute power of God, any body, as such, would have to be extended—or else it would not be a body at all.

In this way, considering God's absolute power also amounts to a withdrawal of the world in which we find ourselves. It allows for the consideration of things outside *this* world but not outside of all worlds. As Ockham argues, God's power is never "inordinate," that is, it is always productive of a world. That God produced *this* world is fundamentally contingent and, therefore, we can suppose its annihilation by considering God's absolute power. This very theological position does, to be sure, appear as such at several crucial points in Hobbes's philosophy. In his debates with Bishop Bramhall, for example, Hobbes argues that God does not will something because it is good, but rather it is good because God wills it.[28] In other words, a notion (presumably rational) of the good does not determine God's will, but rather God's will is the determining ground of the good. Hobbes consistently shows that the assumption of a notion of the good that would determine God's will would thereby place limitations on divine omnipotence. It would also allow a citizen to disobey laws because they violate this good. In his Critique of White's *De Mundo*, Hobbes is concerned that natural theology not overstep its bounds and limit God's omnipotence. Therefore, we must be on guard against the ghosts and spirits of Aristotelian metaphysics.

More than this direct appeal to the late medieval voluntarist tradition, the supposition of the annihilation of the world carries certain features that belong to the theology of God's absolute power—features that bear on the theory of being and knowing. First, the ability to suppose the annihilation of the world and then to suppose a creation or replacement makes our actions (in the supposition, not in reality) similar to the divine. Our will, outside the supposition, depends on appetite and aversion, which in turn depend on sensation, and in this differs from the divine will. God's will, since it is eternal, cannot be thought to be determined or necessitated by anything prior to it, since we can assume nothing prior. The supposed annihilation of the world does not, as we have seen, annihilate memory of what existed before. Yet those memories do not determine our re-creation or replacement. This is what provides a stop in the flow of causation and phenomena—the same pause that the appeal to the absolute power of God provided for medieval thinkers. This stop in the flow of causation and phenomena puts us close to the site that God occupies with regard to our world. This quasidivine site or moment frees our thought from the determination it usually receives from things.

Secondly, and more problematically for Hobbes, the ability, the possibility, of imagining the annihilation of the world does not seem to arise from what Hobbes tells us about our ideas and imagination. Hobbes claims that "phenomena (appearances) are the principles of knowing other things, it must be said that sensation is the principle of knowing those principles and that all science is derived from them" (De Corp. IV, 25, 268). It is clear that the supposition of the annihilation of the world does include memories of things sensed and so might be said to begin in sensation. However, does the knowledge of a universe annihilated begin in sensation? What is more, how do I, in the face of the flow of material things, choose to suppose something so radically different from my sensory experience? How is sensation the principle of that choice? What capacity, if not sensation, opens the possibility of such a choice as to suppose the universe annihilated? It seems clear that this appeal to something analogous to the absolute power of God must be based on a capacity other than sensation, a capacity that does not even have its origin in sensation, and therefore has a ground other than material bodies.

I have already shown that Hobbes's restriction of the scope of philosophy provides the ground of his critique of Aristotelian substantial forms. Yet that restriction stated that the philosopher ought to dispute only those things that are from sensation *or* reason (or, indeed, both). Reason, then, cannot be identical with what appears to the senses and must have an origin outside the senses. Since the supposition of the annihilation of the world brings about a pause in sensation, reason may be just that capacity that allows the supposition in the first place and that derives the results of first philosophy. In this case, the statement "whatever is, is a body" is made not from within, as it were, the material motions of phenomenality but from the pause that reason can bring about in those motions. Yet this raises a second question. How is this appeal to a capacity of reason consistent with materialism, since this capacity needs to be independent of the phenomenality of material bodies?

THE WORK OF REASON

After having shown that Hobbes's materialism is not primarily and originally a metaphysical posit, and after having shown that first philosophy—in which the notion of body operates—stands outside the flow of phenomenality and its causation, one is left wondering what, within Hobbes's philosophy, authorizes this break with material phenomenality. In his critique of substantial forms, Hobbes already indicated that philosophy is bound by what appears to sense and what is available to reason. Such a claim only makes sense if the operation of reason is independent of the flow of phenomena that appears to sense. In other words, if reason is nothing more than a sequence of images, then there is no possibility for anything other than phenomenality to become, as it were, the matter for thought. Such a position, however, presents difficulties concerning Hobbes's arguments that move from natural right to natural law. Without reason performing a task other than what is given it by sensation, Hobbes's political philosophy would be merely descriptive—that is, it would not provide the normative grounds of a legitimate political body.

Reason, in fact, is present already from the very beginning of the *De Corpore* when Hobbes insists that a *ratio naturalis* is innate in all humans (De Corp. I, 1, 11). It occupies a central role in the definition of philosophy, which is the ground upon which the rest of the *De Corpore* operates. "Philosophy is knowledge [*cognitio*] *acquired through correct ratiocination*, of effects or rather phenomena, from the concepts of their causes or generations, and on the other hand of generations, which are possible, from known effects" (De Corp. I, 1, 12; emphasis mine). The qualification that philosophy is knowledge acquired through correct ratiocination allows Hobbes to exclude from philosophy both sensation and memory because, although both are knowledge, they are given immediately and not acquired through ratiocinating (Ibid.). This exclusion does not mean that philosophy has nothing to do with sensation. On the contrary, we have already seen that sensation is one of the two sources of knowledge to which philosophy is bound. Hobbes here denies that sensation *as such* is philosophy precisely because of its immediate givenness.[29] From what is immediately given, one could never legitimately claim "whatever is, is a body," precisely because the concept "body" is not immediately given to sensation.

Ratiocination, therefore, must expose mediation in that it uncovers *how* things are. As such, it must stand, as we have seen, outside the flow of phenomena that are immediately sensed. This mediation emerges because ratiocination computes, that is, "collects the sum at the same time of many things added or knows the residual of something subtracted from another" (Ibid.). This computation can occur even without words, requiring only the idea of something. To show this, Hobbes uses an example that medieval thinkers used to show both the primacy of the concept of "being" and the validity of the famous "tree of Porphyry."[30] If someone sees something obscurely at a distance, that person will have an idea of that thing, which, after having imposed a name, is called body. When later that same person sees the body move from place to place, a new idea occurs, to which the name animated might be imposed.[31] Two crucial features of ratiocination emerge here. First, although Hobbes will later say we add and subtract names, here he shows that naming is by no means essential to ratiocination. Hobbes recasts the traditional example to show how one can add and subtract without the use of names. Indeed, the translation of this image out of its medieval context at the same time effects a break between the knowledge that reason uncovers and the source of that knowledge in the world—that is, it already effects a break in the identitarian thinking Adorno calls "metaphysics." Second, and perhaps more importantly, Hobbes uses this image of obscure vision to show how *new* ideas emerge not just because of sensation, but because of some "work" done on sensation, namely the addition of one idea to another. This act of adding produces a new idea that is other than what is merely sensed. In other words, the collection of ideas produces not merely a combination of ideas that emerge from sensation, but a new idea that is more than the totality of sensations. Thus the work of reason in adding and subtracting works *on* sensation in order to produce knowledge *about* what is sensed while not being bound to the flow of phenomenality. This is precisely the mediation reason provides. The work of reason neither opposes the givenness of what is in favor of identifying it with what is originally a concept belonging to thought nor takes that givenness on its "face," for what it purports to be. Reason has an analytical and critical task to perform, precisely in order to uncover the materiality of what presents itself phenomenally.

The first part of the *De Corpore* begins with the definition of philosophy and ends with a discussion of method. The method, however, unpacks what is already implicit in the definition of philosophy. Philosophy is the knowledge either of effects from the concept of their causes or of the possible causes from a conceived effect. "The method of philosophizing, therefore, is the investigation of effects through their known causes or of causes through known effects" (De Corp. I, VI, 1, 57). This dual direction is found already in Aristotle's *Posterior Analytics*. There he argues that we can have ἐπίστήμη in two ways, though one is more properly ἐπίστήμη than the other. We can either have knowledge of a fact—the planets do not twinkle—or knowledge of the *why*. These are both scientific—ἐπίστήμη—because even the *fact* can be demonstrated. The planets are near, that which is near does not twinkle, therefore the planets do not twinkle. But then again, that which is near does not twinkle, the planets do not twinkle, therefore the planets are near. The difference between these two demonstrations is not just the incidental placement of premises. Rather, one syllogism shows that the reason *why* the planets do not twinkle is *because* they are near. The other does show the fact that the planets do not twinkle, but does not lay out the *reason why*. It is not *because* the planets do not twinkle that they are near, though they are, in fact near and that nearness affects their twinkling. Rather, it is because they are near that they do not twinkle.[32]

This same duality, namely that we can demonstrate a cause from its effect or an effect from its cause, belongs to Hobbes's method. "And thus science is τοῦ διότι or of causes; another is all knowledge of what is called τοῦ ὅτι, is sensation or is imagination or memory remaining from sensation" (De Corp. I, VI, 1, 58). From this, Hobbes draws the conclusion that the principles of all knowledge are phantasms, that is, knowledge begins in sensation. However, there is a difference in the knowledge of what is sensed and the knowledge of the causes of that sensation. The first is given immediately and "naturally." The second, however, "consists in composition and division, i.e., resolution" (De Corp. I. VI, 1, 1, 58). Hobbes concludes from this that every method that deals with causes is "either compositive, resolutive, or partly compositive and partly resolutive" (De Corp. I, VI, 1, 1, 58). When knowledge of what is merely given is involved, no method is required because sense and sensation is sufficient. When, however, knowledge of cause, that is, the reason why, such a thing is given is involved, then a method that engages the capacities of reason is required.

As Hobbes continues his discussion of method, he begins to unpack precisely why a method is required at all. That is, if sensation is the most obvious of all capacities in that it delivers its object immediately, it seems as if no method is required at all. Yet, the difficulty is that sensation delivers a "whole," that which is given without an understanding of what makes it what it is.

> In knowledge of the senses, however, the whole phenomenon is more known than any part of it; when we see a human, first or

> more known is the concept or idea of that whole of human than the particular ideas of *figured, animated, rational,* that is, we see and know the whole human, which is, before we turn the soul to those particulars. (De Corp. I, VI, 1, 2, 58)

The difficulty here is that something like a human is given in sensation and yet that human is nothing if it is not figured, animated, and rational. These "parts" are crucial to an understanding of the human that is given in sensation because without knowledge of these parts we might never know, for example, how that human is moved by other bodies. That motion depends not on its being human but rather on its being a body, a body of a certain figure, and so on. Thus, what is given in sensation, the whole, is not sufficient for an understanding of how that thing is able to be affected and to affect other bodies. For this knowledge, the "work of reason," that is, composition and resolution are required. However, the results of that work are not given immediately in the whole that is what is sensed.

The knowledge of causes, however, that is, of the "parts" of that whole that was sensed, starts, indeed, with those parts. But how do we arrive at knowledge of the parts if only the whole is given in sensation? We have already seen one such strategy in Hobbes's "first philosophy," namely the supposition of the annihilation of the universe. Indeed, such concepts as body, figure, motion, and so forth, were uncovered there in the gesture that withdrew all wholes from sensation. Since sensation presents what Marx called a "chaotic representation of the whole," Hobbes recognizes that the causes of that whole are not sensed with it, except chaotically. So philosophy needs a different method that neither rejects the reality of the material body present to sensation nor clings to its chaos. In short, Hobbes recognizes the dangers of Althusser's "empiricism" in not separating off the real, essential kernel from the merely apparent surface.

There is the danger of ideology on both sides. On the one hand, Marx exposed the danger in accepting as immediate that which is, in fact, mediated. Empiricism and a blunt materialism fall into this danger. It can be called ideological because it stamps what has come to be in a historically determinate way with the label of "real." It does not turn to the capacities of reason in order to uncover the various forms of mediation by means of which that which presents itself to sensation is a whole whose parts do not show themselves. On the other hand, there is the ideology that forms of thought and imagination have escaped from our heads and are pretending to be real.[33] Hobbes's insistence that there must be a method in order to analyze the material body that is given in sensation is a rejection of what Adorno called "identitarian thinking" as much as a reject of the materialism that claims that materiality is immediately given.

According to Hobbes's definition of philosophy, sensation is certainly the principle of all knowledge. It is, however, not philosophy. Up to the closing chapter of Part I of *De Corpore,* Hobbes has only said that philosophy requires the work of reason and, therefore, sensation alone does not count as philosophy. Here, however, Hobbes begins to show why that must be the case. Without knowledge of what makes a given whole, sensation is limited to the givenness of that whole. Philosophy, however, seeks to know [*scire*] the "cause of a certain phenomenon" (De Corp. I, VI, 3, 59). This task, which Hobbes calls "science *simpliciter*," cannot remain with just the cause of this or that phenomenon but must ultimately seek the "knowledge of the causes of all things" (De Corp. I, VI, 4, 59). Upon seeking the cause, for example, of *this* material body, I must ultimately have recourse to the most fundamental causes of all things, namely those concepts that are uncovered in first philosophy. This is because "the causes of all singulars are composed from the causes of universals or simples" (De Corp. I, VI, 4, 59).

I noted above how the universals that are simple and the "causes of all singulars" cannot be entirely the same as universal names. They do, however, share one feature in common: they are both signs. For Hobbes, a sign is similar to what he calls a note or, perhaps, a mark [*nota*], in that both are material "monuments," that is, a material remembrance. Both signs and marks have their origin in the memory some animals have of their forgetfulness. We establish a monument to remember only on the condition that we understand the possibility of forgetting. The difference between the memorial functions of marks and signs is that marks are private while signs are public. One can institute anything to mark and aid in the remembrance. This is seen, for example, in the custom of tying a string around one's finger. The string marks something but that connection between mark and thing is available only to the one instituting it. "The difference between marks and signs is that marks are instituted by us, signs thanks to another" (De Corp. I, 2, 2, 20).

However, if both marks and signs are material monuments to aid memory, how could one (marks) be private while the other is public? The only indication Hobbes gives is in his definition of signs: "Signs are accustomed to be called the antecedents of consequents and the consequents of antecedents, as we often experience those to precede or follow in a similar way" (De Corp. I, II, 2, 20). Here, the cause/effect relationship seems to be what allows signs to be public. It is not private that dense, black clouds are a sign that rain will follow. The causal relationship is available to all. And signs are just this causal relation. Thus there are natural signs, like dark clouds.[34] Signs instituted by choice—literally, arbitrary signs—are also antecedents of consequents or consequents of antecedents. The only difference is that they are willful. When, for example, a person places stones to mark the borders of his land, he wants to show to others something.[35] The antecedent of sign, however, is only that something was in the thought of the one offering the sign (De Corp. I, II, 3, 21). Therefore, a sign is only a consequent of a thought in the sense that it is instituted in order to show to others that one is thinking of something. It is a material mark of that which occurred in thought.

As such, signs are always of concepts, not of things themselves. "[I]n what other sense can the sound of this spoken word 'stone' be understood than that the one who heard it gathers that the speaker had thought of a stone?" (De Corp. I, 2, 2, 21). Since signs have no direct and immediate relation to something that is but only to the thought of the one offering it, the hearer can only assume that the one offering it thought of a stone, a tree, a human, and so on.

This shows that, the "universals contained in the nature of singulars"[36] are not the same universals that are just names of names of things. The universals that are discovered by the philosophical method are "dug out" [*eruenda sit*] by reason and, therefore, are not just names of names. To resolve a given singular, into a "plane terminated by a certain number of equal lines and right angles" (De Corp. I, VI, 4, 59) is not the same as "line" being a sign of many lines taken singly. What is "dug out" by reason is the causes or generations of that very singular given as a whole in sensation. The method of resolution, that is, the digging that reason performs, does not, according to Hobbes, leave us with the incessant arbitrariness of signs but, rather, discovers the causes of that which is given as a whole in sensation. Thus the resolutive method that brings us to the most simple and, therefore, most universal notions, is crucial to philosophy, since philosophy seeks the causes of what is given immediately in sensation.

Hobbes's reference to digging out recalls Althusser's metaphor for the process he called "empiricism." Althusser, however, focuses on one aspect of the metaphor: in mining gold, there is something of value, something essential—the gold—that is contained within a crust, shell, or clump of earth or rock that must be separated. For him, the essence is dragged out of that which is inessential. Hobbes, however, digs not for an essence but for the cause of that which appears. It is, therefore, similar to Marx's move to "ever thinner abstractions" in that the more simple names cover more things.[37]

Hobbes digs but he does not extract. The causes, the simple universals that are discovered by the digging of reason are only the possible causes in the sense that such things could produce the effects from which we began. Recall that Hobbes's definition of philosophy moves in two directions. Philosophy is either knowledge of effects from *known* causes or knowledge of *possible* causes from known effects. The immediacy of sensation provides direct knowledge of effects, but not knowledge of how these effects are produced in us—that is, sensation does not reflect on the mediation by means of which phenomena are sensed by us. The move from sensed effects to knowledge of their causes does not produce knowledge of those causes that, in fact, have produced the thing. The ability of reason to uncover not actual but only possible causes marks that very freedom of reason from phenomenality that was operative in the supposed annihilation of the world. Reason can only generate knowledge of what can produce such effects, even if they have not been produced, in fact, in that way. Such knowledge does not deserve the name "adequate," if by adequate we mean a correspondence between concept and thing. Yet such knowledge is not, therefore, nothing. Such knowledge would allow us to produce the effect, or reproduce the effect, even though the effect as existing was not produced in this way. This, indeed, is the *telos* of philosophy for Hobbes—that one be able to produce effects for the commodious well-being of human life.[38] For such a task, all that is needed is knowledge of how a given effect might be produced (De Corp. I, 1, 14). In this way, the independence of reason from phenomenality also frees knowledge from the chain of efficient causation by which things are, in fact, produced and therefore sensed. The majority of Hobbes's philosophy, therefore, and especially his analysis of sensation, does not state how things really are, but how they might be produced. The knowledge gained of individual bodies, for example, does not authorize the assumption that this knowledge grasps the "real being" of the things known. There is no possibility of asserting an identity between the being of a thing and its being known. This means that the capacity of reason to analyze knowledge must proceed along different lines than that knowledge that it analyzes. This is the site in which Hobbes's materialism emerges in such a way that it is not the mythical, (quasi)theological positing of matter as being—things do not come to be from Night. Here, Hobbes seeks to establish reason and the concepts it uncovers on a ground other than the immediacy of sensation. And yet, Hobbes argues that the concepts of reason do not grasp a reality that stands behind phenomenality. Reason works *on* phenomenality, finding concepts that allow that phenomenality to be traced, without insisting that phenomena find their ultimate reality in a kind of being that does not appear. This work of reason prevents the identitarian thinking that constitutes Aristotelian metaphysics and more emphatically its scholastic reception. This is why Hobbes's materialism is inseparable from his critique of that metaphysics.

Hobbes's definition of philosophy rejects the pure immediacy of phenomena as the proper object of philosophy, thus insisting on the role of mediation through reason. If that mediation, however, would provide adequate knowledge of what immediately appears, he would not be a materialist, but well on the road to idealism. Without attention to mediation, on the other hand, he would be saying little more than "All is from Night." Instead, Hobbes insists that the task of philosophy is to expose the mediation through concepts of reason of what is immediate to sense. The appeal to the mediation of reason, however, does not deny the reality of phenomena as such. The reality of what appears immediately to sensation is left untouched. Reason's task, therefore, is not to get to a "higher level essentiality" of mere appearance, but to present, through the mediation of reason, the way in which phenomena appear. This task is inscribed by him in the larger task of bringing about effects for the commodious well-being of human life.

For Hobbes, the basic task of philosophy is to uncover the causes of things not as a purely theoretical task but as one way to put things to work so that life may be better. The task, therefore, is to bring about *commodum*, advantage, opportunity, a favorable condition—in short, something close to happiness or at least one of its conditions. It is in the name of something like a better condition that Hobbes refuses Aristotelian/scholastic metaphysics. His mining operation uncovers causes so as to put them to work for a condition more favorable to human life. It is only reason, the work of reason, that can dig out the causes of things so that we might use them to bring about a life that is other than the one we are currently living. In this sense, Hobbes's materialism works from the past in the sense that it uncovers causes but toward a future that is designated by nothing more than its being other and better.[39]

This is what makes Hobbes a materialist, much more than his position "whatever is, is matter." Without reason performing its work independently of the material things that appear immediately, matter is in fact not independent of our knowledge of it. Yet the irreducibility of matter to thought is surely the hallmark of materialism. It turns out, therefore, that materialism, if it is to maintain the independence of matter from thought, must allow matter to emerge from the mediation of reason. If Hobbes were to argue that being is matter because only matter appears to a being who senses, he would be arguing that reality must be identical to our knowledge of it. Yet in such a case, matter ceases to have material reality, since its reality would be determined by knowledge. In order to insist on a reality to which our concepts are not adequate, the materialist must focus on the way in which phenomenal immediacy is mediated by reason—a mediation that does not determine the reality of that phenomenal immediacy. This work of reason, identified already at the beginning of the *De Corpore,* is carried out in Hobbes's first philosophy—the proper materialist response to scholastico-Aristotelian metaphysics. This focus on mediation forms the core of Hobbes's critique of substance metaphysics. Yet, as we already saw, the ground of this critique is political. By pointing out that reason grasps only the possible causes of things (except those things of which we, ourselves, are the cause), Hobbes can show that the notion of substantial forms posits more than is legitimate. The insistence on the nonidentity of reason and what is immediately raises the question of what, materially, is at work in the scholastico-Aristotelian claim about substantial forms. Hobbes can then claim that the work of this metaphysics is actually ethical and political, providing a ground for moral claims that justify civil war and regicide. The materialism in Hobbes emerges precisely in the independence of reason from phenomenal immediacy that leads to discovering causes that do not constitute reality as it is, but as it might be made by us for a more favorable life.

REASON AND NORMATIVE LEGITIMACY

The independence of reason, which allows it to provide mediation in the face of the flow of phenomenality, is also crucial in the movement Hobbes makes from natural right to natural law. This move is pivotal in that the natural laws Hobbes uncovers provide the motivation for the covenant that constitutes the sovereign and, thereby, the commonwealth. The independence of reason from phenomenal immediacy—an independence that makes possible the supposition of the annihilation of the universe—also frees reason from the present (sensation) and the past (memory). The future is precisely the site where politics emerges.[40] The natural condition of humans is one of equality of strength and mind. This equality means that there is also an equality of hope in achieving ends. Such an equality of hope leads to diffidence when the same end is desired by more than one person. "And from this diffidence of one another, there is no way for any man to secure himself, so reasonable, as Anticipation" (Lev. XIII, 69). It is this condition that makes life without a common power a condition of war. War, for Hobbes, is not merely an actual conflict but the future intention to use force to secure one's ends. In this way, the entire state of war is based on the conception of the future that each has. No one is presently my enemy who is not actually threatening me. However, I can envision the *desire* in others to enjoy my goods and to use even my body for the attainment of their own ends. My fear, therefore, results from the indeterminacy of the future.[41]

This fear of an indeterminate future can be expressed in the form of a right that belongs to all outside of a common power: each has the liberty "to use his own power, as he will himselfe, for the preservation of his own Nature; that is to say, of his own Life, and consequently, of doing anything, which in his own Judgement, and Reason, hee shall conceive to be the aptest means thereunto" (Lev. XIV, 72). This natural right, however, turns into a radically different "Law of Nature" precisely through the application of reason—reason that must, if it is to lead to this law of nature, be independent of sensation and its constant present. The law of nature demands that everyone seek peace, insofar as there is hope of attaining it. In this way, reason grasps a norm that binds precisely because reason alone is capable of hoping for a future that appears indeterminate in the present and also capable of hoping for norms that would make that future determinate. In other words, the natural right that each has to everything is descriptive rather than normative. The transition from *right*, which is the liberty to do or not to do, to *law*, which determines and binds, is "found out by reason" (Ibid.). If one projects this right of nature into the future, one discovers that, in such a future, one's security and life are always precarious. This projection, which demands a capacity other than sensation, leads to the laws of nature and ultimately to the covenant that constructs the sovereign. A covenant, according to Hobbes, differs from other transferrals of rights in that a covenant points not only to past and present, but also to the future. "The matter, or subject of a covenant, is alwayes something that falleth under deliberation; . . . and is therefore always understood to be something to come; and which is judged possible for him that covenanteth, to performe" (Lev. XIV, 77).

The second transition, from the laws of nature to the commonwealth, arises precisely because the laws of nature are normative (i.e., determining and binding *in the face of* natural right) and not merely descriptive. "For the Lawes of Nature . . . of themselves, without the terrour of some Power, to cause them to be observed, are contrary to our naturall Passions. . . . And Covenants, without the sword, are but words, and of no strength to secure a man at all" (Lev. XVII, 93). Since Hobbes has already traced the passions to sensation, the transition from natural right to natural law and from natural laws to covenant must be one that constitutes a commonwealth outside the flow of phenomenality and the material bodies that cause sensation. This is what gives the natural laws their normative character. And yet, since sensation yields to passions, the natural laws do not, in fact, bind without the threat of power.

This work of reason, the projection of a future from out of the mediation that reason provides in the face of sensation, allows the discovery of the laws of nature, and therein lies the first normative moment of Hobbes's political thought. That these laws do not bind *of themselves* indicates their independence from phenomenal materiality. Yet the move to the covenant is based on that other materialism, the sociopolitical materialism, that I have argued stands at the basis of Hobbes's philosophy. The schoolmen, with their immaterial substances, attempt to ground norms in reality, though a reality that precisely does not show itself to sensation or reason—the spirits that are "blown into men."

Here, however, an interesting question emerges. If reason is able to expose the mediation of what appears immediate, if reason is able to pause the flow of phenomenality, and therefore discover norms, is not reason also able to hope for a different set of natural laws, ones based not merely on self-preservation, but on the very capacity of reason itself to remain independent of sensation and materiality? Such a projection would entail that one protect, in the future, something like the freedom of reason, a freedom that is precisely not protected by the sovereign. The difficulty in producing a normative philosophy on the basis of this freedom of reason, however, is that it would seem to threaten the very materialism that always motivates Hobbes's thought. Yet still, the freedom of reason stands there at the core of Hobbes's philosophy, allowing us to wonder whether the sovereign's authority and power are delivering for us all that we desire, given the conditions of reason.

That reason is, for Hobbes, *both* the condition for digging out how material things are *and* the normative ground that leads to the covenant that establishes sovereignty brings about a certain dialectic—perhaps even the dialectic of Enlightenment itself. What prevents this from spinning into dialectics, however, is Hobbes's insistence on self-preservation as striving that belongs to material beings, and living ones most of all, as such. Yet the very capacity of reason that he requires to "dig out" the causes of things and to see a way out of the war to which natural right naturally tends is the same capacity that leads to a critique of self-preservation—"the world is as dialectical as that."[42]

NOTES

1. The phrase is from Aristotle, *Metaphysics*, trans. Hugh Lawrence-Tancred (London: Penguin Books, 1998), 1072a20. I will return later in this chapter to this theme.
2. Theodor Adorno, *Metaphysik: Begriff und Probleme* (Frankfurt: Suhrkamp, 1998), 14.
3. As I indicated in the Introduction, it was this anomaly that shifted my project away from its basis in Hobbes toward a more general investigation into materialism.
4. See, for example, Leo Strauss, *Political Philosophy of Hobbes: Its Basis and Its Genesis*, trans. Elsa M. Sinclair (Chicago: University of Chicago Press, 1952); Howard Warrender, *Political Philosophy of Hobbes: His Theory of Obligation* (Oxford: Clarendon Press, 1957); as well as Tom Sorell, *Hobbes* (London: Routledge, 1991). This is the central issue in the debate between Sacksteder and Warrender.
5. Thomas Hobbes, *Critique Du De Mundo de Thomas White*, ed. Jean Jacquot and Harold Whitmore Jones (Paris: Vrin-CNRS, 1973), 312. This is also consonant with Hobbes's treatment of body in the *De Corpore*, as I will show below.
6. Ibid.
7. Ibid.
8. Ibid.
9. Aristotle, *Metaphysics*, 1071b. See also *Metaphysics* 1072a19.
10. Thomas Hobbes, *Leviathan*, ed. Richard E. Flathman and David Johnston (New York: Norton, 1997), 215. Hereinafter this text will be cited as Lev., followed by chapter number and the page reference to this edition. It should be noted that the premise of this entire section of *Leviathan* is the reality of ghosts, spirits, and demons.
11. In this way, Hobbes's "materialism" is not all that different in method from Marx's. As argued above, Marx's materialism does not shy away from

conceptual generality precisely because it is based on a sociopolitico-economic critique.

12. For most of his life, Hobbes was better known as a natural philosopher than as a political philosopher. Mersenne, for example, included a text of Hobbes on optics in his *Universae geometriae synopsis*. Descartes also first knew of Hobbes as a natural philosopher. One might argue against making any methodological claims based on the structure *De Corpore* since, although it was the first volume projected for Hobbes's *Elements of Philosophy*, it was actually written after the third volume, *De Cive*. However, Hobbes himself reinforces his original methodological plan while breaking it by publishing *De Cive* first. In the *Preface* to *De Cive*, Hobbes reasserts the proper method, beginning with first philosophy and the rest of *De Corpore*, moving on to the human in *De Homine*, and culminating with *De Cive*. However, because of the civil war in England, he felt compelled to publish *De Cive*. While waiting for the proper methodological ground, *De Cive* can stand on its own because its principles are known by experience.

13. I will investigate Hobbes's concept of reason, which is defined by him as addition and subtraction, in part 4 below.

14. This is a phrase that Hobbes uses frequently. It stems, I would argue, from his desire to take seriously the middle voice of the Greek φαίνεσθαι.

15. Hobbes, *De Corpore*, ed. Karl Schuhmann (Paris: J. Vrin, 1999), 75. Hereinafter, this text will be referred to as De Corp. followed by Part, chapter, and page reference to this edition.

16. Ibid.

17. Yves Charles Zarka, *La Décision Métaphysique de Hobbes* (Paris: J. Vrin, 1999,) 40.

18. Rene Descartes, *Meditationes de Prima Philosophia*, VII:17. Oeuvres de Descartes (Paris: J. Vrin, 1983). Unless otherwise noted, all translations from this text are my own.

19. Ibid., VII:172; the emphasis is mine.

20. Ibid., VII:173–74.

21. I discuss ratiocination in more detail in the next section.

22. Hobbes uses "image" and "phantasm" interchangeably throughout his works.

23. In the previous chapter, we saw Scotus's solution to this problem in his argument that a species or common nature is *real* even if not existing or actual. Although Hobbes is here a nominalist in the tradition of Ockham, I argue that he comes close to Scotus when it comes to the most general or common causes that are also real even if not existing.

24. We should note here the decisive difference between Aristotle's insistence that whatever is must be something and Hobbes's claim that whatever is must be simply whatever.

25. Zarka, *La Décision Métaphysique de Hobbes*, 40.

26. For a discussion of the role that this distinction plays in Ockham, see my *Science, the Singular, and the Question of Theology* (New York: Palgrave, 2002); in Pierre d'Ailly, Richard Lee, *Force of Reason and the Logic of Force* (Hampshire: Palgrave Macmillan, 2004), 85–98. For the relation between this tradition and Hobbes, see Zarka, *La Décision Métaphysique de Hobbes*, 46ff.

27. To the extent that political theology originally indicated the fact that the principles and concepts of political theory are secularized theological concepts and principles, political theology starts with this basic insight of later medieval thinkers. Medieval theologians recognized that if God has the power to make laws, including and especially the regular order of nature that makes the world

intelligible, then the current order of law is *but one possibility*. If God is *all-powerful*, that is, if God can do or make *anything* (whether, as in Scotus and Ockham, bound by the principle of noncontradiction, or, as in d'Ailly, not bound by that law), then the initial possibilities open to divine power must be conceived as infinite. These possibilities were conceived as possible orders, that is, orders of law. As such, the absolute power of God stands outside any order and, as a consequence, outside of any lawfulness. This is not a "state of exception" precisely because an exception is made *to a law*. This is what drives d'Ailly to acknowledge the possibility that God is not "bound" by the principle of noncontradiction and also that God is best conceived not as the first cause, but as the first law-giver. Pierre d'Ailly, *Quaestiones Super I, III et IV Sententiarum* (Paris: Jean Petit, 1499). On this issue of the absolute power of God and the law, see also Carl Schmitt, *Politische Theologie. Vier Kapitel zur Lehre von der Souveränität*, Auflage: 9., Aufl. (Berlin: Duncker and Humblot, 2009) and Giorgio Agamben, *Homo Sacer: Il Potere Sovrano E La Nuda Vita* (Torino: Einaudi, 1995).

28. Thomas Hobbes, *Questions Concerning Liberty, Necessity, and Chance*, ed. William Molesworth, *English Works of Thomas Hobbes of Malmesbury*, vol. 5 (Darmstadt: Scientia Verlag Aalen, 1966). Such a position also entails that behind the divine will there is nothing that determines God's creation, not even reason.

29. Once again, the problem with the immediacy of what is given comes to the fore. As I have argued, it is immediacy that leads to both the critique of materialism (e.g., Aristotle's critique of Thales) and to Marx's difficult materialism that tries to show that what is immediate is constructed and, therefore, never as immediate as it seems.

30. Peter of Spain, *Summulae Logicales*, ed. I. M. Bochénski (Turin: Marietti, 1947).

31. The concept of "imposition" was central to the medieval logic known as "terminist," because it attended to the fact that terms have characteristics that affect the logic of a proposition and, ultimately, a syllogism. Perhaps the most influential logician to highlight the importance of "imposition," was Roger Bacon. See Bacon, "De Signis." *Traditio* 34 (1978): 75-136.

32. This discussion is found in *Posterior Analytics*, Book I, chapter 13.

33. Karl Marx and Friedrich Engels, *Collected Works of Karl Marx and Friedrich Engels, 1845–47, Vol. 5* (London: Lawrence and Wishart, 1975), 5:23.

34. Here, Hobbes does not differ at all from late medieval, terminist, and primarily nominalist logicians like Ockham. This definition of signs and the subsequent distinction between natural and arbitrary signs was already put forward by Roger Bacon. See Bacon, "De Signis."

35. This kind of metaphor extends throughout Hobbes's philosophy. On this question, see Michael Bray, "The Hedges That Are Set," Epoché 11 (2006): 173–200.

36. De Corp. I, VI, 4, 59.

37. The term *eruo* is used, for example, in relation to gold mining and pulling a plant out by the roots. Therefore, it also can mean "to draw out," much like "to abstract." Finally, as "destroying by removing the foundations," it shares certain features with deconstruction.

38. At De Corp., I,1,6,14, Hobbes defines the end [*finis*] or scope [*scopus*] of philosophy as "that we can use previously seen effects for our convenience [*commoda*], or by the application of bodies to one another, we might produce effects like those conceived by the soul, to the extent that human strength and industry and the matter of bodies will enable, for the well-being of human life."

39. In the following chapter, I expose this same dynamic in Benjamin and Adorno.

40. In a way, all of philosophy is directed toward the future, since its *telos* is the application of knowledge of cause to bodies so as to bring about effects in the future for commodious well-being. This futural aspect of philosophy is made possible, as I have argued, by Hobbes's insistence that philosophy always include *ratiocination*. I argue in chapter 5 that this might be better thought as "it could be otherwise," rather than as a future that could always succumb to utopian nightmares.

41. Once again see Bray, "The Hedges That Are Set."

42. Theodor Adorno, *Negative Dialektik: Jargon Der Eigentlichkeit*, vol. 6, Gesammelte Schriften (Frankfurt am Main: Suhrkamp, 1970), 51; Theodor Adorno, *Negative Dialectics*, trans. E. B. Ashton (New York: Continuum, 1973), 41.

FIVE

Benjamin and Adorno on the Social Need for Materialism

Already in 1951, Adorno linked his thinking of what, in *Negative Dialectics,* he called materialist metaphysics, to the notion of redemption [*Erlösung*]: "Philosophy, as it is responsible in the sight of despair, would be the attempt so to treat all things as they would be shown from the standpoint of redemption."[1] Why does Adorno here insist on the seemingly theological category of redemption? What stands in need of being made right? And furthermore, does not redemption present a flight, if not into the transcendent, at least out of the world in which we find ourselves? And if redemption presents a flight, is it not precisely a flight away from the materiality of the world in which we find ourselves?

At the end [*zum Ende*], Adorno turns to the notion of redemption, with all its messianic overtones. At the end, there is only one [*einzig*] philosophy that can face despair, namely the philosophy that "exposes the cracks and rifts" in which the world is set, the cracks and rifts that would be laid out "in the messianic light."[2] The cracks and rifts to which Adorno here refers open up precisely because, as we have seen, thought sets out to think that which is and yet goes errant, misses that which was its intended "object." In so doing, thought posits its own conditions in place of what is. This gesture was that which Althusser called "empiricism." Yet, as Althusser argued, the conditions for this mode of thinking are not, themselves, native to thought—they are actual social conditions. As a result, what is posited as what is turns out to be nothing but the legitimization of those very social conditions. The conditions become the standard not just of thought but of what is. In relation to those conditions now posited as the norm, the world appears alien. This, in short, is the rift. The actual world is experienced as other than, alien to, alienated from, the world that is posited by the conditions of thought.

The task of thinking in the light of redemption, however, is not straightforward. This is a result of the implicit link between the standpoint of redemption and transcendence; for the standpoint of redemption is a perspective that "set[s] the world beside itself, alienated from itself" and, as such, is also impossible "because it presupposes a standpoint at a remove, were it even the tiniest bit, from the influence [*Bannkreis*] of existence."[3] In this way, we are once again walking between two errant paths: on the one hand is the path that posits the transcendence of thought as the actual reality of what is, on the other hand is the path that insists on direct and unmediated access to what is. Each path is traveled only when one is under a spell, perhaps even haunted.

The attempt to trace this path between two errant wanderings, however, is fraught. "[M]eanwhile any possible knowledge must not only be wrested from that which is in order to be taken as binding, but for that same reason smacks of the same disfigurement and neediness that it intends to escape."[4] The standpoint of redemption sees the world as disfigured and needy, as it will appear in a messianic light. However, the standpoint also presupposes a point of remove, even the smallest distance, from the actual world. The standpoint of redemption, therefore, attempts to pry free from that which is another possibility. In so doing, it risks becoming itself disfigured and needy. At a remove from the actual world, which is what thought was to think, thought now becomes a caricature of itself and in need of an other to think. As a result, thought falls back into the world without being able to think what is from the standpoint of redemption. In such a situation, it simply puts its seal of approval on the world as it is given: "the real world is as it ought to be."[5] Yet in order to avoid the catastrophe of authorizing what is just as it is, philosophy, Adorno argues, must risk thinking in the light of redemption, for there is a demand imposed on it to think what is in light of another possibility, even at the tiniest remove. "Yet over-against the demand that is issued to it, the question of the reality or non-reality of redemption is itself almost indifferent."[6]

In his 1965 lecture course, *Metaphysics: Concept and Problems*, Adorno indicates that metaphysics always moves in two directions.

This duality can be seen most clearly in Aristotle, the first metaphysician on Adorno's reading. One direction of Aristotle's metaphysics is its critique of the Platonic insistence that only ideas have being, while actually existing things are deficient copies of the ideal.[7] As such, Adorno argues, the factical appears as nonbeing in relation to the being of ideas. The "first thesis" of Aristotelian metaphysics is "that the universal is not substantial; that it is not, like the Platonic Ideas, which are universal concepts, something existing in and for itself, independently of its realization."[8] In this sense, Aristotle's metaphysics is a critique of Platonic thought.[9] On the other hand, for Aristotle universal concepts are "not a *mere* abstraction in relation to the particulars subsumed under [them]."[10] This dual movement belongs to all metaphysics: "while these systems have always been critically disposed towards anything they regarded as dogmatic or fixed ideas, they have attempted, on the other hand, to rescue [*erretten*], on the basis of thought alone, that to which the dogmatic or transcendent ideas referred."[11] Aristotle presents not only a critique of the Platonic notion of the Idea, he also attempts to save what it was that Plato's thought tried to say by means of the Ideas.

This notion of rescue [*Rettung*] is not far from Adorno's insistence that philosophy always think in light of redemption. In fact, the last section of *Negative Dialectics,* "Reflections on Metaphysics," shows precisely the need to think through this dichotomy: in the light of Auschwitz,[12] metaphysics—as long as it clings to a notion of the best of all possible worlds or a cosmic plan—has made a mockery of our experience and yet, in the face of that experience, to say "this is all there is," makes a mockery of the suffering that has been experienced.

> One has the feeling that the depth of philosophical reflection, which is necessary as a resistance to all the illusion with which reified consciousness surrounds us, at the same time leads away from the truth, since one sometimes suspects that this same existence which it is the inalienable impulse of philosophy to penetrate and go beyond, is the only thing which exists and is worth reflecting on *at all.*[13]

Critique and rescue are both needed so that philosophy can make good on the promise of metaphysics that it failed to achieve.

Adorno's argument, however, is not merely that the characteristic physical suffering of the twentieth and twenty-first century must be taken as "given" or "immediate," and, as such, must fundamentally alter philosophical and metaphysical reflection, though that is true. Rather, his argument sets out from what I have been calling Aristotle's "errant path." In his notes that formed the basis of the third lecture, May 18, 1965, Adorno writes, "The entwinement of metaphysics with *thought*, inaugurated so emphatically by Aristotle in opposing hylozoism, is irrevocable."[14] For Adorno,[15] Aristotle "at the beginning" insists that only the actually existing, individual thing counts as a being, and yet "universal concepts are not mere abbreviations of the particulars subsumed under them—rather they have an attribute which raises them above *flatus vocis*, above the mere breath of the voice."[16]

At the beginning, and here Adorno refers to Aristotle's *Categories*, what counts as being, as what is, is that which "needs nothing else—which requires no means [*was keines Mittels bedarf*] by which we can perceive it; it is, in other words, the *immediate* [*das* Unmittelbare]."[17] Yet this very position concerning the fundamental reality of the immediate flies against what Adorno sees as the main, and indeed positive, feature of metaphysics first brought about by Aristotle: it is a thinking of mediation. This is the basic feature of Aristotelian metaphysics for Adorno. It is a philosophy of mediation that fails to think that very mediation. The evidence of this is Aristotle's notion of μορφή. For Aristotle, whenever there is talk of "form," there is always a "reference to something of which the entity in question is the 'form.'"[18] The other term, matter, ὕλη, "cannot be conceived as something which does also not have form."[19] Therefore, there is a basic relationship between these two—they are mediated—and yet Aristotle does not think that mediation. This, for Adorno, constitutes the errancy of Aristotle's metaphysics.

This errancy is related to that which Althusser pointed to as "empiricism" in that there is a process whose result is taken as if there were no work to produce it.[20] The process of bringing about the result is precisely what Adorno here calls mediation. The lack of mediation is apparent in Aristotle's linking, if not outright identification, of form as the "force immanent in ὕλη, matter, this concept of the immanent idea which at the same time is the power centre which moves matter."[21] When form comes into close proximity with actuality, with ἐνέργεια, matter becomes the not yet formed. As such, according to Adorno, although matter is never without form, there is still a priority of form over matter just as there is a priority of actuality over potentiality, of ἐνέργεια over δύναμις. In this way, as I argued in my reading of Aristotle,[22] form becomes the higher reality because it, as such, is realized, while matter is only on its way to being formed, to being realized. This leads Adorno to point out a fundamental tension that characterizes Aristotle's metaphysics: on the one hand, he maintains that the only reality is what is immediate, the "this something," τόδε τι, on the other hand, he insists that the "mediated, the concept realized and formed in matter," is the higher reality.

Adorno, however, is not attempting to criticize Aristotle for falling into an inconsistency. Rather, he sees in Aristotle's errancy the fundamental problem of materiality: "But you may be able to recognize here something of the 'misère de la philosophie', in the fact that even this non-conceptual element [matter], this non-principle, which nevertheless is constitutive of, and inherent in, all philosophy, cannot appear within that realm—which, heaven knows, can only operate with concepts—except in the form of a concept."[23] Matter is both required for thought and other than thought and, therefore, can only be brought into the realm of philosophy by means of a concept. Yet as a concept, matter is no longer matter—its materiality is stripped away. It is this fundamental insight that motivates the entire project of *Negative Dialectics*.

As Adorno points out, "dialectics" is the name of a method almost as old as philosophy in the West. It initially had a positive spin as a way of approaching accepted and received truths in order to overturn, or at least critique, them. Yet in Kant's philosophy, dialectic emerges only when something has gone wrong with our rational capacities. We are led into contradiction by not recognizing that reason has gone beyond the bounds that it has set for itself. In this sense, at least on Adorno's reading, Kant subscribes to a form of the traditional notion that truth cannot contradict truth. Therefore, when thought, when reason, comes to a dialectic, that is, to a moment when two contradictory truths are uncovered, this is a sure sign that all is not well with thought. In Hegel's hands, however, that Kantian moment is given a slightly different spin. It just may be the case that we are forced to admit that the very condition for the possibility of a truth is its contradictory opposite. This is not a sign of failure but an engine for reason, for thought, to push forward to a truth that contains both of the previous, contradictory truths.

For Adorno, dialects, "says no more nor no less than that objects do not rise up into [*aufgehen*] their concepts, that they come into contradiction with the traditional norm of *adequatio* . . . [Contradiction] is the index of the untruth of identity, of the rising up of the conceived into the concept."[24] In this way, Adorno sees contradiction at the very core of what it means to think, what it means to engage in the attempt to grasp not just "that it is" but to attempt to understand the "why." There is, therefore, a seemingly basic recognition that stands at the heart of Adorno's project: philosophy, "heaven knows," also operates with concepts but concepts are always concepts *of* . . . . The "of" indicates that what is conceived is necessarily other than the concept of it—even if what is conceived is another concept. "In truth all concepts, even the philosophical ones, go out to non-conceivables, because they are on their part moments of reality, which are necessary for their construction—primarily for the goal of the domination of nature."[25]

There is, therefore, always a reference to a "something" that cannot be denied. This is the thesis of nonidentity. Phrased in this way, however, we seem to be on the terrain of those philosophies that claim that there is "stuff," objects, things, rocks, and so on, "out there" with which philosophy ought to begin. Yet Adorno recognizes that this is still the Aristotelian dream of a first philosophy.

> The critique of ontology sets its sights on no other ontology, even a non-ontology. That would instead merely place another as the absolutely first; this time not absolute identity, being, the concept, but the non-identical, beings, facticity. It would thereby hypostisize the concept of the non-conceivable and operate adversely to what it intends. A philosophy of grounds [Grundphilosophie], πρώτη φιλοσοφία necessarily leads to the primacy of the concept with itself; what refuses [the concept] also leaves behind the form of an alleged philosophizing from out of the grounds.[26]

In this way, we are, it seems, caught between the Scylla of the critique of metaphysics on the basis of its errancy and the Charybdis of the dream of a ground in the nonconceivable. As Adorno argues here, the basic insight that thought requires an other that, necessarily, will not be identical to it means that that other will always resist being conceived. "To think is, in its proper sense, to think of something. Even in the logical form of abstraction of something, as a meant or a judged, which claims, of itself, to be based on no being, lives the indelible of that non-identical for thinking, which it would like to delete, that which is not thought."[27]

Such a situation is precisely that which leads to dialectics. The "truth" of the concept depends on its opposite, the nonconceptual. Adorno, however, is here demanding a dialectics that is negative in the sense that it does not achieve some "higher level" in which this contradiction is resolved. On the one hand, the claim to immediacy reasserts first philosophy and claims a ground in that which is. This is still a claim to a ground and, therefore, is still a claim that what is can, after all, be thought. On the other hand, the claim of thought is always the claim that its abstractions are real—Marx's ghostlike objectivities have gotten out of our heads. There is, therefore, a demand that philosophy engage in a critique of thought and its need for concepts *from out of the position of thought itself.*

While Adorno does not make it explicit, negative dialectics is demanded because, as Marx pointed out, the danger of ideology is not countered by another ideology of immediate access to an object. A philosophy oriented toward an object must, at the same time, recognize its own conditions. The thing, an object, is, Adorno argues, prior to any subject for two reasons. First, as I have argued, the thought, the unique property of the subject, is always the thought *of something* that is necessarily other than that thought. Therefore, the thinking subject depends on an object in a way that the object does not depend on the subject. Second, as Kant already saw in the second antinomy, every subject is also an object, while every object is not also a subject. There is, therefore, a priority of the object over the subject. However, this priority does not mean that "objectivity would be an immediacy [*ein Unmittelbares*], that the critique of naïve realism would be forgotten. Priority of the object means the progressive, qualitative differentiation [*Unterscheidung*] of the mediated in itself, not of that beyond, but itself articulated in [dialectics]."[28] The thing, the object other than thought, as Scotus argued, is ultimately given priority if it is thought as irreducible differentiation.

If I could think whatever I wanted, dialectics would not be what any right-thinking subject would choose. Dialectics is no ontology. It is akin to a fever in that it is a potentially self-protective expression of life that, at the same time, could be fatal. As Adorno insists, "dialectics is the ontology of the false condition."[29] It is appropriate to a specific condition, one that is brought about by non- or extra-philosophical conditions. Dialectics is the only way in which thought can grasp that what does not exist, "ghost-like objectivities," are nonetheless effective and, therefore, real. When ghosts operate in the world, how can we think this mode of operation? Adorno's answer is that dialectics is forced upon thought for this very reason. The wrong condition emerges, for Adorno, when the positive capacity of thought turns against itself, when, as Marx diagnosed in *The German Ideology*, what belongs to thought has escaped out of our heads and now is out "there" doing actual work—

abstractions are operative, functional, effective. We are confronted with a duplicity in "objects." The first object, that in which we want to invest a fundamental existence and, therefore, priority, is that of which thought is supposed to be about. The other is the ghost, the abstraction now walking in the world and, as much as philosophically we want to say "does not exist," operates and has effects.

This is the fundamental difficulty with any claim to "immediacy." Yes, the object has priority, and yet that priority cannot mean that we grasp the object, the thing, immediately in such a way that the claim to immediacy can be the *Grund* from out of which thought sets. The impasse Adorno indicates in the passage quoted above is that the demands of thinking philosophically are only satisfied by means of concepts. This prevents any insistence that objects are the first, the basis, the ground from out of which philosophy ought to proceed. Yet, concepts themselves, as we have seen, are dependent on that of which they are concepts—they always lead out of themselves to that which is necessarily other. On one side, the claim to an immediate access to that which is other than the concept, to the object, is the fundamental rejection of philosophical analysis. On the other side, to insist that concepts are the "higher level essentiality" of that of which they are concepts is to block the very hope of concepts themselves, the hope to be concepts *of something*.

Adorno's method, therefore, begins not from the object but from the concept and its need for it other. For him, however, this means turning the basic philosophical question away from "why" toward "how." The question of how emerges in two interrelated directions. First, there is the original insight of the Hegelian dialects that contradiction is no longer ontologically determinative. That is, to point out a contradiction is not, for that reason alone, to point out that which is not real. This leads to the question of how that which is a contradiction can be. Second, there is the question of how that which is can be grasped by a thinking that is, by definition, other than what is. The question of "how" is, for Adorno, always the question of mediation. That what is can only be grasped through a process or mechanism that is necessarily other always means that what is necessarily is also mediated through something that is external to it. There is an irreducible distance between what is and the thought of it. But that distance is marked in that there is a trace of the object that, even while mediated, is nonetheless part of the very mediation that thought engages. Therefore, negative dialectics sets out from that trace, from the mark of that which is other than thought, in order to recognize the "priority of the object."[30]

This issue of mediation is not unrelated to the "ghost-like objectivity" that returns over and over again whenever philosophy sets out to think matter. The moment that Marx uncovers the objectivity of ghosts is the moment that the principle of non-contradiction ceases to have ontological validity. Yet that is the same moment in relation to which the question "why" must give way to the question "how"—how can that which cannot exist nevertheless be real in the sense that it counts as an objectivity? This question forces philosophy to look outside its own norms for truth and validity. How can that which, by the traditional norms of philosophy, is false still be real?

It is this question that turns Adorno's thought toward the social as the only possible way to understand that reality of "ghost-like objectivities." *Negative Dialectics* centers on the structural isomorphism between what Adorno calls "identity-philosophy" and Marx's insight into exchange as the origin of value in capitalist societies. As I argued in chapter 2, Marx attempts to think through how material things such as iron and corn and coats, all of which presents themselves as things independent of thought, can ever be taken as equivalent so as to be exchanged. Their presentation as material things, that is, in those qualities and characteristics that make them the things they are, does not allow them to be exchanged because, as the particular material things they are, they have no feature on the basis of which they could be seen as equal so as to be exchanged. In short, nothing about their actual existence forms the real basis for their exchange. And yet, they are exchanged and, therefore, identical, at least in some respect.[31]

What Adorno sees is that this identity between two things that are not identical is isomorphic with the gesture, first seen in Aristotle, that identifies thought with that which is not identical to it—thought is seen as identical to the "object," to the material thing that thought is supposed to be "about." In each case, there is an identity between two poles that are not, actually, identical. This condition can be put almost formally: A is not A, yet A is A because they are exchanged as equivalent; or the thought (idea, essence) of A is not A, yet the thought of A is A. The difficulty is that the logic presents a contradiction and yet it is, if not true, nonetheless real. At this point, right thinking wants to insist that the contradiction ends the question: A cannot equal, cannot be, not A. Yet in both cases it turns out that A really is not A. If the traditional norms of philosophy fall short here, then perhaps those norms are inadequate. However, the traditional norms are still helpful here. The recognition, on the basis of the traditional norms, that what is a contradiction cannot be, along with the clear reality that what is contradictory nevertheless *is*, leads us to strive to think how what cannot be (what is false) still is real. Such a situation calls out for that form of philosophy that will not stop once it reaches a contradiction, namely dialectics.

Yet simply to say "dialectics" will not be sufficient because we are looking for a condition that cannot be philosophical in origin, since the philosophical norm at the same time brings it about and forbids it. So we must search for a condition under which what is and what is not can be identical. That condition is, Adorno argues, social because it is the condition under which ghostlike objectivities operate. Negative dialectics, therefore, sees the principle that Marx uncovered, the principle that underlies exchange, as central to the critical task of philosophy. Just as a principle of identity "grounds" the possibility of exchange, so too a principle of identity grounds the philosophy that takes what belongs to thought as the being, the essence, the "higher level essentiality" of that which is other than thought. The condition for the possibility of exchange and the condition for the possibility of extracting the essence away from the material thing are the same. As a result, the critique of the gesture of identifying what is with what is thinkable in it is always also a social critique.

> [S]ocial motifs—in [Aristotle's] case the powerlessness of the individual, the retreat into private life, all the moments we call Hellenistic—do not influence thought from the outside, but that, in a way which is difficult to pin down and which has not been analysed in principle up to now, they impinge on the immanent coherence of thought itself, making themselves felt in the internal argumentation of the individual philosopher. If a theoretician of intellectual history were to attempt to understand this highly remarkable connection, which establishes within philosophy a logic which is in curious harmony with social experiences imposed from outside, without that philosophy having to adapt itself externally to them, that would be an extremely important task.[32]

The general characterization of metaphysics as identity-philosophy, that is, as the positing of the concept as the being of the thing conceived, also carries along with it the notion that the meaning of what is, of the factical, is only found in the transcendent, in another world. Metaphysics as the identity of thought and being is, therefore, inherently antimaterialist as it turns away from what is given toward what is thought. This antimaterialist metaphysics, however, flies in the face of what Adorno calls "today." What must be refused today is the insistence that only the immutable "is true and substantial while the transient is inferior and despicable, a mere mode or deception of the senses, as which it has been tirelessly denounced by philosophers since Plato."[33] The refusal to insist "on a positive meaning or purpose in being" emerges, Adorno argues, after Auschwitz.

> And as a *sign* of this—the word symbol would be wretchedly inadequate, since we are concerned with the most symbolic thing of all—I will take Auschwitz. Through Auschwitz—and by that I mean not only Auschwitz but the world of torture, which has continued to exist after Auschwitz and of which we are receiving the most horrifying reports from Vietnam—through all this the concept of metaphysics has been changed to its innermost core.[34]

That change to its innermost core [*bis ins Innerste*] is that metaphysics now has materialism thrust upon it. After Auschwitz, Adorno refuses the identification of being with thought because that identification no longer makes sense out of the world in which Auschwitz occurs. However, after Auschwitz, Adorno also refuses to give up metaphysics entirely. The traditional gesture of metaphysics, namely to expropriate meaningful existence from what is to that which belongs to thought can only say to suffering that it has a purpose, a meaning. This gesture makes a mockery of suffering. However, that same gesture also says, "This is not all there is, it could be otherwise." The way to think the critique of metaphysics while keeping its promise is precisely that metaphysics that has materialism thrust upon it.

The "Meditations on Metaphysics," with which *Negative Dialectics* ends, sets out from the most basic insight of the text: concepts always are constituted by an "other" that is lost the moment the concept is posited as identical to its object. The principle of identity that characterizes both the society built around exchange and the metaphysics that denies the materiality of matter is a principle of infinite and absolute exchangeability, fungibility. The experiences Adorno indicates under the sign "Auschwitz" are experiences of precisely that absolute fungibility. Just as ten yards of linen *is* one coat—the *Wertgegenständlichkeit*—so too every human *is* any other. The individual, the subject, is not immune from the results of identity-philosophy and that identity-philosophy has a social form whose ultimate expression is Auschwitz.

Yet metaphysics also promises an "otherwise" in the face of what is and in this promise there is something that Adorno variously characterizes as happiness and hope. The idea of happiness belongs, Adorno argues, precisely to the Scotistic notion of the irreducible singularity that is called *haecceitas*. "Only in the face of the absolutely, irreducibley [*unauflöslich*] individuated can it be hoped that exactly this has been and will be; to fulfill this would fulfill the concept of the concept. But the concept clings to the promise of happiness, while the world, which denies it, is that of dominating universality."[35] It is this definition of happiness, namely the irreducible individuality of what is, that opens itself onto that which is other than the concept. Singularity, however, is not the opposite of the concept—except in that metaphysical gesture that, having extracted what belongs to thought from the material thing, now calls that extracted essence the reality of the thing from which it was extracted. Yet on the other side, even in the metaphysics that Althusser called empiricism, in identity-philosophy, the other necessarily remains. "In the objectivity of metaphysical categories, not only is the solidified society condensed . . . but all the more so the priority of the object as a moment of the dialectic."[36] The categories of metaphysics, as I have argued, are dependent on an other in a way that the other is not dependent on those categories. A concept is always a concept *of something* and, therefore, that something not only is prior to the concept but is also that which the concept strives to think. Adorno's critique of metaphysics as identity-philosophy has shown this. The flip side, however, is that the promise, the striving, of the concept exposes the danger of its liquidation: it regresses "to the subjectivism of the pure act. It hypostasized the mediated as immediate."[37] Happiness, therefore, demands both the recognition of irreducible singularity—*haecceitas*—and the rejection of the pure act of rampant subjectivism. It demands a self-critique of thought *so that what is not identical to the concept can emerge in its nonidentity*.

Adorno's materialism, therefore, is not the turn toward the "nonidentical" outside of thought—as if thought had immediate access to it. The ruthless critique of thought is precisely what allows that which is not identical to the concept to emerge in its difference. This is what Scotus has taught us. But the need for this is made all the more urgent "after Auschwitz." That is, what is required is not the rejection of metaphysics but the opening onto a metaphysics that has materialism thrust upon it. The insistence on metaphysics, indeed the insistence of metaphysics, stems from the dual source of its own inherent promise—on which it has historically failed to deliver—and the appropriate response to the suffering of the victims of all of those ever expanding experiences gathered under the sign "Auschwitz." This is why, already in *Minima Moralia*, Adorno insisted that a philosophy that does not think in the light of redemption is not worth thinking.

The relation of happiness to redemption as materialism is one of the central features of Benjamin's "On the Concept of History."[38] Historical materialism is already announced in the first fragment of this text. Indeed, several of the beginning fragments end with an odd sort of intervention of this "historical materialism." For example, Fragment II, which deals precisely with the connection of happiness and redemption, ends with the claim, "The historical materialist knows this."[39] Fragment IV, a reading of the fundamental character of class struggle, ends with "The historical materialist must be aware of this most inconspicuous of all transformations."[40] Over the next several fragments, Benjamin returns over and over again to what the historical materialist must know, do, understand, and think.

Returning to Fragment II, Benjamin links happiness immediately to redemption: "There is happiness—such as could arouse envy in us—only in the air we have breathed, among people we could have talked to, women who could have given themselves to us. In other words, the idea of happiness is indissolubly bound up with the idea of redemption."[41] The progression of tenses and moods needs to be marked here. There is happiness only in relation to the past. Furthermore, that past—"the air we *have breathed*," "the people we could *have talked* to"—is sometimes, though not always, also marked by the subjunctive—we *could have*. The idea [*Vorstellung*] of happiness is therefore marked, if not constituted, by a relation to the past but also by an other or otherwise. *There is* [*es gibt*] happiness only in relation to the past and the past is also marked by an "otherwise," by what *could* have been.

It is, therefore, no wonder that what has been said of the idea of happiness is also said of the idea [*Vorstellung*] of the past. If happiness is linked to the past and the past is linked to redemption, then the past is, by some form of the commutative property, also linked to redemption: "The past carries with it a secret index by which it is referred to redemption."[42] Perhaps happiness is linked to redemption because it is linked to the past and the past is linked to redemption. But what about the other of the past or present, the otherwise? Benjamin recoups this: "For does not a hint of the air which has been for the past not touch us also? Is there not in the voices that are given to our ear an echo of the now silent? Do not the women who we come on to have sisters who they no longer recognize?"[43] The otherwise that could have been is not something that belongs to the past because it is constitutive of us here, now. The air we breathe, the voices we hear, contain within themselves both a past and, in direct relation to that past, a future that is at once the future of that past and *only one* future of that past. It is the "only one" in relation to the past that requires the subjunctive. There is an asymmetry as we trace back from the present to the past and as the past necessarily projects an otherwise. This asymmetry stems precisely from the "otherwise."[44]

As Benjamin ends the fragment, the link back to redemption becomes central precisely on the basis of this asymmetry. "In that case, that we are on the earth was awaited [*erwartet worden*]. Then, we, as every generation [*Geschlecht*] that was before us, have been given a *weak* messianic power, on which the past has a claim. Such a claim is not to be made good on cheaply."[45] There is here an important issue that relates to the difference between expected and awaited.[46] The past comes to light here as shot through with possibility in the sense that it is always open toward a future. As Benjamin argued from the earliest essays of his writing career, that openness to the future is precisely what makes a mechanistic understanding of time[47]—one that underlies the relation of cause and effect—impossible. To be sure, the present is the future *of the past*. Therefore, a line can be drawn from the present back to the past. However, seen from the other direction, there is not one line but a multiplicity that is, perhaps, infinite. Each moment, therefore, is always just this point from which, perhaps, an infinite number of lines can be drawn. We, I, am but a point on but one of those lines. While I have a direct connect to *the* past, I was only awaited by that past.

In this way, the entire fragment thinks through the asymmetry I raised above. It is because of this asymmetry between the present and the past that the past can make a claim on the present in a way that the present cannot make on the past. What the past makes claim to is the "*weak* messianic power" that we have, *because*, Benjamin insists, our being on the earth was awaited. Here again we have a relation between past and present that is asymmetric. The fact that our being on the earth was awaited means that the past is a condition for our being. When, however, I trace the conditioned, that is, my being on the earth, back to its condition, the past, that is when I find any past moment as shot through with perhaps infinite moments, any one of which could be the condition seems to be unconditional. The traditional, at least since Kant, way to express this would be to say that what has a condition and therefore is conditioned, that I am on the earth, has a condition that, since shot through with infinite—or at least indeterminate—moments that could be future, must be unconditioned. That about which Benjamin seems most insistent here is that this fundamental asymmetry is the basis for *both* a claim the past has on me *and* a power I have, however weak, to redeem that past.

Benjamin ends Fragment II with two odd claims. The first is that "This claim is not to be settled cheaply [*billig*]."[48] The claim that the past has on the present now appears to be something like a responsibility. If the claim could be settled cheaply, then simply being on the earth might suffice to make good on it. However, that seems insufficient, as that price is, indeed, cheap. More must be done to make good on this claim. Yet what is this more? What more is to be done? A first indication emerges precisely from the asymmetry I have been tracing between the present and the past. If the past is shot through with infinite lines *any one of which could be the future,* and if my being on the earth is *but one of those lines,* I must now make good on the infinite, the unconditioned, or at the very least indeterminate nature of that "waiting for" of the past that is the condition for my being on the earth. To pay off a debt that is perhaps infinite is certainly not going to be cheap.[49] I will return to the question of what more can be done to settle this claim below.

The end of the fragment, however, is perhaps the most perplexing: "The historical materialist knows of this."[50] This is the second mention of historical materialism, the first coming in Fragment I. Here, however, materialism, historical materialism, is tied directly to the weak messianic power we have to redeem the past. As in Adorno, we have the claim that philosophy must think in the light of redemption and that thinking is called "materialist." What is materialist about this claim of the past and our messianic power to redeem it? At first, we should note the origin of both the claim and the power to redeem it: our being on the earth was awaited. What is the sign of this waiting for?[51] What is the sign of the openness of the past toward an otherwise?

Adorno ended *Negative Dialectics* with exposing the need "today," that is, in light of those experiences of events for which the proper name "Auschwitz" cannot be enough, to do justice to both the horrifying recognition of what is and also the promise of metaphysics that there is an otherwise. In doing so, he returns to this

Benjaminian insight of the claim and the power of redemption. Both Adorno and Benjamin call this claim-redemption relation "materialism" without any reference at all to matter and the material conditions of what is. Or rather, they turn to materialism precisely in order to show that what is, what is other than thought, as a condition of "today," of the present, does not immediately present itself to us. What Andrew Benjamin says, in commenting on Walter Benjamin's *On the Concept of History*, of the past, holds just as well for the present: "What comes to be repeated throughout the fragments that comprise *On the Concept of History* is a concern with what the 'authentic' (*eigentlich*) image of the past actually is. Where what 'it' is, is not 'it' as it appears."[52] The materiality of what is, we have learned from Marx, Scotus, Hobbes, Adorno, and now Benjamin, is not as it appears and, as such, materialism must encounter the ghosts that walk around as if they were things.

The relation between the redemptive, the messianic, and materialism turns precisely on this insight. Materialism is, in fact, the only way to exercise our weak messianic power to make good on the claims of the past. However, it turns out that in order to redeem the past—indeed in order to understand the "it is" of the past—is to also try to find a way to understand that "it" is not as it appears. Materialism, if it is to be the faithful thought of what is, therefore, necessarily requires—if it is to think "in the light of redemption"—a way of thinking that which is other than thought that, at the same time, recognizes the "other" that presents itself to thought. That recognition does not come by means of an insistence on that other. The nonidentical, *haecceitas*, is both other than thought and that toward which thought strives. But to insist on it as the ground also collapses the hope to think that which is. In order to think what is, we *must* think that which cannot be, yet operates. We have to think about ghosts. To think what is against the ghosts is what Scotus, Marx, Althusser, Adorno, and Benjamin call materialism.

And so we have to end where we began, with my aunt Peggy.[53] This is, on my reading, the insight of Benjamin's *On the Concept of History*. I pay off too cheaply the claim that she, that the past, has on me if I simply say that there is stuff about which philosophy has nothing to say but which, nonetheless, makes a claim on thought. If we are to think that which is, and if that which is always is a thing that is constituted by materiality, then we have to recognize that the thinking of matter runs counter to materiality. The central dilemma of the thought of matter comes to a head in just this moment. It is the *thought* of matter that constitutes materialism. Adorno's insight,

Benjamin's insight, is that this means that thought cannot be rejected, all the while it strives to make good on the *of*, of the openness, the otherwise, that I have argued belongs to matter in what it is, in its materiality.

Earlier, in my discussion of the final section of Adorno's *Negative Dialectics*, I insisted that Adorno thinks the promise of metaphysics as happiness and as hope. At that moment, I analyzed his claim only in the direction of happiness. But we are now drawn back to the question of hope precisely because it is the correlate to our messianic power to redeem. I cannot speak more clearly and directly about this than to return to my aunt Peggy's (re)introduction of me at my aunt Betty's funereal. That I am on the earth was awaited. Yet that cannot explain Aunt Peggy's pride in my being a professor *of philosophy*.[54] Only Benjamin's insistence on the asymmetry of the relation of the present moment to a past moment can help to understand this. The "what it is" of a life in all its material conditions also includes the living, the going on, that, as its own condition, is shot through with an opening to a future that is not conditioned. That I am her godson does not settle the claim she has on me. That I am a professor goes a bit further, but for her still did not settle the claim. That I am a professor *of philosophy* seemed to her to go further in settling the claim that the past has on me. In this introduction, Aunt Peggy provided a way to understand the proper sense of materialism. The *thought* of what is, the *thought of matter*, requires not an attentiveness to what is—of that we, after "Auschwitz," are all too keenly aware. What the thought of matter requires is attentiveness to the otherwise, to the other than thought that no claim to an immediate relation to what is can provide.

In the end, there is no end of ways in which we have access to what is in all its difference from our thinking. However, the claim that the materiality of what is, as such, presents a challenge to that very thinking is offensive to what Benjamin and Adorno have shown to be the redemptive power of thought—a redemptive power that is more deserving of the name "materialism" than any other pretender to the throne. What materialism must be today and at the end is an insistence that justice is possible. That justice cannot, however, renounce the capacities of thought that can think through the ghosts that walk around as if they are things.

NOTES

1. Theodor Adorno, *Minima Moralia: Reflexionen aus dem beschädigten Leben,* vol. 4, Gesammelte Schriften (Frankfurt am Main: Suhrkamp, 1951), 283; Theodor W. Adorno, *Minima Moralia: Reflections on a Damaged Life* (London: Verso, 2005), 247.

2. Adorno, *Minima Moralia: Reflexionen aus dem beschädigten Leben,* 4:283; Adorno, *Minima Moralia,* 247.

3. Adorno, *Minima Moralia: Reflexionen aus dem beschädigten Leben,* 4:283; Adorno, *Minima Moralia,* 247. The term *Bannkreis,* which I have translated as "influence" might better be rendered as "under the spell." The term *Bann,* "spell," is one that appears quite frequently in Adorno's work, particularly in both *Negative Dialectics* and *Dialectic of Enlightenment.* From the context, it is clear that Adorno is concerned about a philosophical method that would claim that what is simply and directly presents itself to thought. As such, existence casts a spell on thought. This could be read as Adorno's interpretation of Marx's materialist method. I was first shown the importance of the use of *Bann* in *Negative Dialectics* by Christopher Turner, a student in a seminar I taught at DePaul University on Adorno's text. I am indebted to his careful tracing of that term in *Negative Dialectics.*

4. Adorno, *Minima Moralia: Reflexionen aus dem beschädigten Leben,* 4:283; Adorno, *Minima Moralia,* 247.

5. The reference is to §39 of Hegel's *Vorlesungen über die Geschichte der Philosophie.* Georg Hegel, Wilhelm Friedrich, Pierre Garniron, and Walter Jaeschke, *Vorlesungen über die Geschichte der Philosophie* (Hamburg: F. Meiner Verlag, 1986).

6. Adorno, *Minima Moralia: Reflexionen aus dem beschädigten Leben,* 4:283; Adorno, *Minima Moralia,* 247.

7. See chapter 3 for a fuller discussion of this issue.

8. Theodor Adorno, *Metaphysik: Begriff und Probleme,* vol. 1, Nachgelassene Schriften: Vorlesungen (Frankfurt: Suhrkamp, 1998), 42–43; Theodor Adorno, *Metaphysics: Concept and Problems,* ed. Rolf Tiedmann, trans. Edmund Jephcott (Stanford: Stanford University Press, 2001), 25.

9. It should be noted that Adorno argues that Aristotle is the first metaphysician precisely because Plato does not think the relation between the transcendent ideas and actually existing things—metaphysics, on his reading, is always tied to the mediation of the transcendent and the factical.

10. Adorno, *Metaphysik: Begriff und Probleme,* 1:43; Adorno, *Metaphysics: Concept and Problems,* 25.

11. Adorno, *Metaphysik: Begriff und Probleme,* 1:19; Adorno, *Metaphysics: Concept and Problems,* 8.

12. In *Metaphysics: Concept and Problems,* Adorno indicates that "Auschwitz" stands for "the world of torture which has continued to exist after Auschwitz and of which we are receiving the most horrifying reports from Vietnam." Adorno, *Metaphysik: Begriff und Probleme,* 1:160; Adorno, *Metaphysics: Concept and Problems,* 101. Later in the course, Adorno adds the atomic bomb to the ever-increasing list of what "Auschwitz" stands for.

13. Adorno, *Metaphysik: Begriff und Probleme,* 1:178; Adorno, *Metaphysics: Concept and Problems,* 114.

14. Adorno, *Metaphysik: Begriff und Probleme,* 1:27; Adorno, *Metaphysics: Concept and Problems,* 14.

15. In the lectures, Adorno indicates that his reading is that of Eduard Zeller, *Die Philosophie der Griechen in ihrer geschichtlichen Entwicklung* (Hildesheim: G. Olms, 1963).

16. Adorno, *Metaphysik: Begriff und Probleme,* 1:44; Adorno, *Metaphysics: Concept and Problems,* 26.

17. Adorno, *Metaphysik: Begriff und Probleme,* 1:48; Adorno, *Metaphysics: Concept and Problems,* 28–29.

18. Adorno, *Metaphysik: Begriff und Probleme,* 1:58; Adorno, *Metaphysics: Concept and Problems,* 35–36.

19. Adorno, *Metaphysik: Begriff und Probleme,* 1:58; Adorno, *Metaphysics: Concept and Problems,* 36.

20. See my earlier discussion of Althusser in chapter 2.

21. Adorno, *Metaphysik: Begriff und Probleme,* 1:58; Adorno, *Metaphysics: Concept and Problems,* 36.

22. See chapter 1.

23. Adorno, *Metaphysik: Begriff und Probleme,* 1:69; Adorno, *Metaphysics: Concept and Problems,* 43.

24. Theodor Adorno, *Negative Dialektik,* vol. 6, Gesammelte Schriften (Frankfurt am Main: Suhrkamp, 1970), 16–17; Theodor Adorno, *Negative Dialectics,* trans. E. B. Ashton (New York: Continuum, 1973), 5. Unless otherwise noted, all translations from this text are my own. Adorno here uses the term *aufgehen,* which I have translated as "rise up into." Ashton adds an additional element to his translation, "without remainder." The difficulty with this rendering is that Adorno's target is what he refers to as *Identitätsphilosophie,* that is, identity-philosophy or the philosophy of identity. For Adorno, this is the philosophy that posits that thought and that of which thought is are identical. If the conceived is simply not identical to the concept of it, then there is not really a question of a "remainder," as if the concept almost gets everything of its object but leaves a little something behind. Rather, Adorno's position is that the concept is not the thing conceived. In this way, perhaps *all* of the thing conceived is left behind or, rather, the concept of it is simply other than what it is in itself.

25. Adorno, *Negative Dialektik,* 6:23; Adorno, *Negative Dialectics,* 11.

26. Adorno, *Negative Dialektik,* 6:140; Adorno, *Negative Dialectics,* 136. Here Adorno is attempting to think through what a philosophy based on giving reasons, as in Leibniz's *der Satz vom Grund,* must engage in relation to that ground. Ashton here, it seems to me, entirely misses the point of the critique. On this, Adorno is working through issues that Heidegger also engages throughout his career, for example in Martin Heidegger, "Vom Wesens des Grundes," in *Wegmarken,* vol. Band 9, Abt. 1, Gesamtausgabe (Frankfurt: Vittorio Klostermann, 1976), 123–75; Martin Heidegger, "On the Essence of Ground," in *Pathmarks,* trans. William McNeill (Cambridge: Cambridge University Press, 1998), 97–135; Martin Heidegger, *Gesamtausgabe. der Satz vom Grund Bd. 10. Abt. 1 Bd. 10. Abt. 1,* 1997; Martin Heidegger, *The Principle of Reason* (Bloomington: Indiana University Press, 1991); Martin Heidegger, *Metaphysische Anfangsgründe der Logik im Ausgang von Leibniz* (Frankfurt am Main: V. Klostermann, 1978); Martin Heidegger and Michael Heim, *The Metaphysical Foundations of Logic* (Bloomington: Indiana University Press, 1984).

27. Adorno, *Negative Dialektik,* 6:44; Adorno, *Negative Dialectics,* 34.

28. Adorno, *Negative Dialektik,* 6:185; Adorno, *Negative Dialectics,* 184.

29. Adorno, *Negative Dialektik,* 6:22; Adorno, *Negative Dialectics,* 11.

30. This is why Scotus is crucial to the understanding of materialism. As I argued, for him the singularizing difference, as the determination of the determinable, always is what the determinable, the common nature, was the essence

or quiddity *of*. A nature is nothing if not the nature of something, and that thing will always be a singular whose singular existence remains outside of thought.

31. The connection between exchange value and identity-philosophy is more difficult to see in Ashton's translation as he consistently renders *Tausch*, exchange, as "barter."

32. Adorno, *Metaphysik: Begriff und Probleme*, 1:154; Adorno, *Metaphysics: Concept and Problems*, 98.

33. Adorno, *Metaphysik: Begriff und Probleme*, 1:158; Adorno, *Metaphysics: Concept and Problems*, 100.

34. Adorno, *Metaphysik: Begriff und Probleme*, 1:160; Adorno, *Metaphysics: Concept and Problems*, 101. In 1965, the United States escalated its involvement in the war between North and South Vietnam. This escalation included troops on the ground but also a bombing campaign that was, perhaps, the precursor to the "strategy" of "shock and awe" that would be used in the U.S. invasion of Iraq. It was in 1965, the year Adorno gave this course, that the protest movement gained strength in the United States and elsewhere.

35. Adorno, *Negative Dialektik*, 6:366–37; Adorno, *Negative Dialectics*, 374.

36. Adorno, *Negative Dialektik*, 6:367; Adorno, *Negative Dialectics*, 374.

37. Adorno, *Negative Dialektik*, 6:367; Adorno, *Negative Dialectics*, 374.

38. Walter Benjamin, "Über den Begriff der Geschichte," vol. I.2, Gesammelte Schriften (Frankfurt: Suhrkamp, 1974), 693–704; Walter Benjamin, "On the Concept of History," in *Walter Benjamin: Selected Writings*, trans. Edmund Jephcott, vol. 4 (Cambridge, MA: Harvard University Press, 2006).

39. Benjamin, "Über den Begriff der Geschichte," 694; Benjamin, "On the Concept of History," 390.

40. Benjamin, "Über den Begriff der Geschichte," 695; Benjamin, "On the Concept of History," 390.

41. Benjamin, "Über den Begriff der Geschichte," 693; Benjamin, "On the Concept of History," 389.

42. Benjamin, "Über den Begriff der Geschichte," 693; Benjamin, "On the Concept of History," 390.

43. Benjamin, "Über den Begriff der Geschichte," 693–94; Benjamin, "On the Concept of History," 390. Translation modified.

44. While I prefer "otherwise," this point is also made by Andrew Benjamin, *Working with Walter Benjamin: Recovering a Political Philosophy* (Edinburgh University Press, 2013), 167: "Despite the unrelenting nature of time's passing, the passing of time which needs to be equated here with the naturalisation of the work of time within historicism, time is not so determined that it has to preclude, by definition, that which could have been other. While it may not have been named, what is central is the always present possibility of what has already been referred to as othering."

45. Benjamin, "Über den Begriff der Geschichte," 694; Benjamin, "On the Concept of History," 390. Translation modified.

46. To be sure, *erwarten* can be translated equally as "expected" and "awaited." My choice of translation, therefore, as all translations, is already a way of understanding, an interpretation. As I hope to make clear here, my argument is that Benjamin does not mean that, for each of us, just our very being on the earth was determinately expected. That is, I read in "to expect" a reference to an intentional object that is determined: I expect just this and nothing else. I see in "to await" an openness in the sense that that for which I am awaiting is not determined, or not nearly so, as what is expected.

47. On the issue of messianic time, see my "The Negative History of the Moment of Possibility: Walter Benjamin and the Coming of the Messiah," in

Rethinking the Frankfurt School (Albany: State University of New York Press, 2002).

48. Benjamin, "Über den Begriff der Geschichte," 694; Benjamin, "On the Concept of History," 390.

49. No wonder that Anselm came to understand the second person of the Trinity—*logos*, God become human—on precisely these economic terms. Anselm, *Cur Deus Homo*, ed. Franciscus Salesius Schmitt, O.S.B, vol. 2, 6 vols., Opera Omnia (Edinburgh: Thomas Nelson and Sons, 1940). Anselm's gambit is that since humans have incurred the debt, only a human can repay it. Since, however, the debt is infinite since it is owed to the unconditioned, only God can settle it. God become human is Anselm's solution to this same problem thought in terms of Christian theology. Here, Benjamin implicitly refuses the Christological solution, perhaps because it is too cheap.

50. Benjamin, "Über den Begriff der Geschichte," 694; Benjamin, "On the Concept of History," 390. "Der historische Materialist weiß darum."

51. Once again, I am forced into a use of language that is well practiced in German and in Chicago, ending a sentence with a preposition. The prohibition now can be seen in its fully problematic nature. "Waiting for" is as much as a "doing" as waiting. The demand that there be an object for which I am waiting is independent of the activity of waiting for. English could say this at least as well as German until the prohibition was imposed from a language practice completely alien to English and its Germanic origins. Chicagoans know about this.

52. Benjamin, *Working with Walter Benjamin*, 170.

53. See the Introduction.

54. I discuss this in the Introduction.

Bibliography

Aaron, R. I. "A Possible Early Draft of Hobbes' De Corpore." *Mind* 54, no. 216 (1945): 342–56.

Adorno, Theodor. *Metaphysics: Concept and Problems*. Edited by Rolf Tiedmann. Translated by Edmund Jephcott. Stanford: Stanford University Press, 2001.

———. *Metaphysik: Begriff Und Probleme*. Vol. 1. Nachgelassene Schriften: Vorlesungen. Frankfurt: Suhrkamp, 1998.

———. *Minima Moralia: Reflections on a Damaged Life*. London: Verso, 2005.

———. *Minima Moralia: Reflexionen aus dem beschädigten Leben*. Vol. 4. Gesammelte Schriften. Frankfurt am Main: Suhrkamp, 1951.

———. *Negative Dialectics*. Translated by E. B Ashton. New York: Continuum, 1973.

———. *Negative Dialektik: Jargon der Eigentlichkeit*. Vol. 6. Gesammelte Schriften. Frankfurt am Main: Suhrkamp, 1970.

Agamben, Giorgio. *Homo Sacer: Il Potere Sovrano e la Nuda Vita*. Torino: Einaudi, 1995.

Ailly, Pierre d'. *Quaestiones Super I, III et IV Sententiarum*. Paris: Jean Petit, 1499.

Althusser, Louis. *For Marx*. New York: Pantheon Books, 1969.

———. *Lire Le Capital*. Paris: François Maspero, 1973.

———. *Reading Capital*. New York: Verso, 2006.

———. *The Future Lasts Forever*. Edited by Oliver Corpet and Yann Moulier Boutang. Translated by Richard Veasey. New York: The New Press, 1993.

Anselm. *Cur Deus Homo*. Edited by Franciscus Salesius Schmitt, O.S.B. Vol. 2. 6 vols. Opera Omnia. Edinburgh: Thomas Nelson and Sons, 1940.

Aquinas, Thomas. *De Ente et Essentia*. Vol. 43. Opera Omnia. Rome: St. Thomas Aquinas Foundation, 1976.

———. *Quaestiones Disputatae de Veritate*. Accessed October 21, 2014. http://www.corpusthomisticum.org/qdv02.html#52314.

Aristotle. *De Anima*. Translated by Hippocrates Apostle. Grinnell, IA: Peripatetic Press, 1981.

———.*Metaphysics*. Translated by Hugh Lawson-Tancred. London: Penguin Books, 1998.

———. *Nichomachean Ethics*. Translated by David Ross. Oxford: Oxford University Press, 1984.

———. *Posterior Analytics*. Translated by Hippocrates Apostle. Grinell, IA: Peripatetic Press, 1981.

Aristotle, Hippocrates. *Aristotle's Physics*. Grinnell Iowa: Peripatetic Press, 1980.

Aristotle, W. *Organon 1949*. Rev. ed. Oxford, 1949. http://www.worldcat.org/title/organon-1949/oclc/613893016&referer=brief_results.

Bacon, Roger. "De Signis." *Traditio* 34 (1978): 75-136.

Benjamin, Andrew. *Working with Walter Benjamin: Recovering a Political Philosophy*. Edinburgh: Edinburgh University Press, 2013.

Benjamin, Walter. *Illuminations*. Translated by Harry Zohn. New York: Schocken Books, 1968.

———. "On the Concept of History." In *Walter Benjamin: Selected Writings*. Translated by Edmund Jephcott. Vol. 4. Cambridge, MA: Harvard University Press, 2006.

———. "Über Den Begriff Der Geschichte," I.2:693–704. Gesammelte Schriften. Frankfurt: Suhrkamp, 1974.

Bernstein, Richard J. *The New Constellation: Ethical-Political Horizons of Modernity/Postmodernity*. 1st MIT Press edition. Cambridge, MA: The MIT Press, 1992.

Bird, A. "Squaring the Circle: Hobbes on Philosophy and Geometry." *Journal of the History of Ideas* 57, no. 2 (1996): 217–31.

Boler, John F. *Charles Peirce and Scholastic Realism, a Study of Peirce's Relation to John Duns Scotus*. Seattle: University of Washington Press, 1963.

Brandon, Eric. "Hobbes and the Imitation of God." *Inquiry* 44 (2001): 223–26.

Bray, Michael. "The Hedges that Are Set." *Epoché* 11 (2006): 173–200.

Bredekamp, Horst. "From Walter Benjamin to Carl Schmitt, via Thomas Hobbes." *Critical Inquiry* 25, no. 2 (1999): 247–66.

Brennan, Teresa, and Carole Pateman. "'Mere Auxiliaries to the Commonwealth': Women and the Origins of Liberalism." *Political Studies* 27, no. 2 (1979): 183–200.

Brown, K. C. "Hobbes's Grounds for Belief in a Deity." *Philosophy* 37, no. 142 (1962): 336–44.

Catania, Francis. "John Duns Scotus on *Ens Infinitum*." *American Catholic Philosophical Quarterly* 67 (1993): 37–54.

Charlton, William. "Prime Matter: A Rejoinder." *Phronesis* 28, no. 2 (January 1, 1983): 197–211.

Damrosch, Jr. "Hobbes as Reformation Theologian: Implications of the Free Will Controversy." *Journal of the History of Ideas* 40, no. 3 (1979): 339–52.

Deleuze, Gilles. *Difference and Repetition*. Translated by Paul Patton. New York: Columbia University Press, 1994.

———. *Différence et Repetition*. Paris: Presses Universitaires de France, 1968.

Derrida, Jacques. *Specters of Marx: The State of the Debt, the Work of Mourning, and the New International*. Translated by Peggy Kamuf. New York: Routledge, 1994.

———. *Spectres de Marx*. Paris: Éditions Galilée, 1993.

———. *Speech and Phenomena: And Other Essays on Husserl's Theory of Signs*. Evanston, IL: Northwestern University Press, 1973.

Descartes, Rene. *Meditationes de Prima Philosophia*. Vol. 7. Oeuvres de Descartes. Paris: J. Vrin, 1983.

Dewan, Lawrence. "*Obiectum*: Notes on the Invention of a Word." *Archives d'Histoire Doctrinale et Littéraire de Moyen Age* 56 (1981): 37–96.

Duncan, Stewart. "Hobbes's Materialism in the Early 1640s." *British Journal for the History of Philosophy* 13, no. 3 (2005): 437–48. doi:10.1080/09608780500157171.

Engel, S. Morris. "Analogy and Equivocation in Hobbes." *Philosophy* 37, no. 142 (1962): 326–35.

Fritsche, Johannes. "Genus and *To Ti En Einai* (Essence) in Aristotle and Socrates." *Graduate Faculty Philosophy Journal* 19/20, no. 2/1 (n.d.): 163–202.

———. *Methode Und Beweisziel Im Ersten Buch Der "Physikvorlesung" Des Aristoteles*. Frankfurt: Verlag Anton Hain, 1986.

Frost, S. "Hobbes and the Matter of Self-Consciousness." *Political Theory* 33, no. 4 (2005): 495.

Frost, Samantha. "Hobbes Out of Bounds." *Political Theory* 32, no. 2 (2004): 257–73.

Gauthier, David. "Hobbes: The Laws of Nature." *Pacific Philosophical Quarterly* 82, no. 3 and 4 (September 2001): 258–84.

Geismann, G. "Spinoza—Beyond Hobbes and Rousseau." *Journal of the History of Ideas* 52, no. 1 (1991): 35–53.

Gert, B. "Hobbes, Mechanism, and Egoism." *The Philosophical Quarterly* 15, no. 61 (1965): 341–49.

Gert, Bernard. "Hobbes on Reason." *Pacific Philosophical Quarterly* 82, no. 3 and 4 (September 2001): 243–57.

Giuffrida, Pietro. "Obiecta, Opposita, Antikeimena Nel De Anima Di Aristotele." *Giornale Di Metafisica* 2 (2013).

Gray, Robert. "Hobbes' System and His Early Philosophical Views." *Journal of the History of Ideas* 39, no. 2 (1978): 199–215.

Hegel, Georg Wilhelm Friedrich. *The Philosophy of Fine Arts*. Translated by F. P. B. Osmaston. Vol IV. London: G. Bell and Sons, 1920.

———. *Vorlesungen über die Ästhetik*. Vol. III. Frankfurt am Main: Surhkamp, 1970.

Hegel, Georg Wilhelm Friedrich, Pierre Garniron, and Walter Jaeschke. *Vorlesungen über die Geschichte der Philosophie*. Hamburg: F. Meiner Verlag, 1986.

Hegel, Georg Wilhelm Friedrich, Charles Hegel, and J. Sibree. *The Philosophy of History*, Mineola NY: Dover Publications, 2013.

Heidegger, Martin. *Gesamtausgabe. Der Satz vom Grund Bd. 10. Abt. 1 Bd. 10. Abt. 1*, 1997.

———. *Gesamtausgabe. Wegmarken Bd. 9. Abt. 1 Bd. 9. Abt. 1*, 1976.

———. *Metaphysische Anfangsgründe der Logik im Ausgang von Leibniz*. Frankfurt am Main: V. Klostermann, 1978.

———. "On the Essence of Ground." In *Pathmarks*, translated by William McNeill, 97–135. Cambridge: Cambridge University Press, 1998.

———. *Pathmarks*. Edited by William McNeil. Cambridge and New York: Cambridge University Press, 1998.

———. *The Principle of Reason*. Bloomington: Indiana University Press, 1991.

———. "Vom Wesen des Grundes." In *Wegmarken*, Band 9:, Abt. 1:123–75. Gesamtausgabe. Frankfurt: Vittorio Klostermann, 1976.

———. *Vorträge und Aufsätze*. Frankfurt am Main: Vittorio Klostermann, 2000.

Heidegger, Martin, and Michael Heim. *The Metaphysical Foundations of Logic*. Bloomington: Indiana University Press, 1984.

Hobbes, Thomas. *Critique du De Mundo de Thomas White*. Edited by Jean Jacquot and Harold Whitmore Jones. Paris: Librairie Philosophique J. Vrin, 1973.

———. *De Corpore*. Edited by Karl Schuhmann. Paris: J. Vrin, 1999.

———. *De Homine De Cive*. Edited by William Molesworth. Vol. 2. Oper Philosophica Quae Latine Scripsit. London: Joannem Bohn, 1839.

———. *English Works of Thomas Hobbes of Malmesbury*. Edited by William Molesworth. Vol. 4. London: Joannem Bohn, 1840.

———. *Leviathan*. Edited by Richard E. Flatman and David Johnston. New York: Norton, 1997.

———. *Questions Concerning Liberty, Necessity, and Chance*. Edited by William Malesworth. In *English Works of Thomas Hobbes of Malmesbury*, Vol. 5. Darmstadt: Scientia Verlag Aalen, 1966.

Hungerland, Isabel C., and George R. Vick. "Hobbes's Theory of Signification." *Journal of the History of Philosophy* 11, no. 4 (1973): 459–82.

Jacquot, J. "Notes on an Unpublished Work of Thomas Hobbes." *Notes and Records of the Royal Society of London* 9, no. 2 (1952): 188–95.

Jesseph, Douglas M. "Galileo, Hobbes, and the Book of Nature." *Perspectives on Science* 12, no. 2 (2004): 191–212.

———. "Hobbes's Atheism." *Midwest Studies in Philosophy* 26 (2002): 140–67.

Jessup, Bertram E. "Relation of Hobbes's Metaphysics to His Theory of Value." *Ethics* 58, no. 3 (1948): 209–17.

Kaminski, Stanislaw. "Hobbesa Teoria Definicji." *Studia Logica* 7 (1958): 43–69.

King, Peter. "Duns Scotus: Ordinatio 2 D. 3 P. 1 Qq. 5–6," 1987. http://individual.utoronto.ca/pking/translations/SCOTUS.Ord2d3p2q5-6.trns.pdf.

———. "Duns Scotus: Ordinatio II D. 3 p.1 q.1," 1987. http://individual.utoronto.ca/pking/translations/SCOTUS.Ord2d3p2q1.trns.pdf.

Kodalle, Klaus-M. "Sprache und Bewusstsein bei Thomas Hobbes." *Zeitschrift Fuer Philosophische Forschung* 25, no. 3 (1971): 345–71.

Lee, Jr., Richard A. "The Negative History of the Moment of Possibility: Walter Benjamin and the Coming of the Messiah." In *Rethinking the Frankfurt School*. Albany: State University of New York Press, 2002.

———.*Science, the Singular, and the Question of Theology*. New York: Palgrave, 2002.

———. *Force of Reason and the Logic of Force*. Hampshire: Palgrave Macmillan, 2004.

Lee, Richard A., and Christopher Long. "Nous and Logos in Aristotle." *Freiburger Zeitschrift für Philosophie und Theologie*. 54, no. 3 (2007): 348.

Leijenhorst, C. "Jesuit Concepts of *Spatium Imaginarium* and Thomas Hobbes's Doctrine of Space." *Early Science and Medicine* 1, no. 3 (1996): 355–80.

Leshen, J. "Reason and Perception in Hobbes: An Inconsistency." *Nous* 19, no. 3 (1985): 429–37.

Lyons, William. "Against an Orthodox Interpretation of Hobbes." *Philosophical Quarterly* 27, no. 109 (1977): 302–12.

Marrone, Steven. "Henry of Ghent and Duns Scotus on the Knowledge of Being." *Speculum* 63, no. 1 (1988): 22–57.

Marx, Karl. *Capital*. Translated by Ben Fowkes. Middlesex: Penguin Books, 1976.

———. *Collected Works [of] Karl Marx [and] Frederick Engels*. London: Lawrence and Wishart, 1975.

———. *Das Kapital*. Vol. 23. Karl Marx Fredrich Engess Werke. Berlin: Dietz Verlag, 1968.

———. *Einleitung Zur Kritik Der Politischen Ökonomie*. Vol. 13. Marx Engels Werke. Berlin: Dietz Verlag, 1961.

———. "Further Selections from the Literary Remains of Karl Marx." *Labour Monthly* 5, no. 2 (August 1923).

———. *Grundrisse*. Vol. 42. Marx Engels Werke. Berlin: Dietz Verlag, n.d.

———. *Grundrisse: Foundations of the Critique of Political Economy (rough Draft)*. Translated by Martin Nicolaus. London: Penguin, 1993.

———. *Zur Kritik der Politischen Oekonomie*. Vol. 13.7. Karl Marx/Friedrich Engels Werke. Berlin: Dietz Verlag, 1971.

Marx, Karl, and Freidrich Engels. *Die deutsche Ideologie*. Vol. 3. Karl Marx Fredrich Engles Werke. Berlin: Dietz Verlag, 1969.

———. *Theses on Feuerbach, The German Ideology and Related Manuscripts*. Reprint. Vol. 5. Collected Works of Karl Marx and Friedrich Engels, 1845–47. New York: International Publishers, 1976.

Mintz, Samuel I. "Galileo, Hobbes, and the Circle of Perfection." *Isis* 43, no. 2 (2009): 98–100.

Naas, Michael. "History's Remains: Of Memory, Mourning, and the Event." *Research in Phenomenology* 33 (2003): 75–96.

Patellis, Ioli. "Hobbes on Explanation and Understanding." *Journal of the History of Ideas* 62, no. 3 (2001): 445–62.

Peirce, C. S. *Collected Papers of Charles Sanders Peirce*. 8 vols. Cambridge: Harvard University Press, 1931.

Peirce, Charles. *The Essential Peirce: Selected Philosophical Writings*. Edited by Nathan Houser and Christine Kloesel. Vol. 1. Bloomington: Indiana University Press, 1992.

Peter of Spain. *Summulae Logicales*. Edited by I. M. Bachénski. Turin: Marietta, 1947.

Probst, Siegmund. "Infinity and Creation: The Origin of the Controversy between Thomas Hobbes and the Savilian Professors Seth Ward and John Wallis." *The British Journal for the History of Science* 26, no. 3 (1993): 271–79.

Raylor, Timorthin. "Hobbes, Payne, and 'A Short Tract on First Principles.'" *The Historical Journal* 44, no. 1 (2001): 29–58.

Read, James H. "Thomas Hobbes: Power in the State of Nature, Power in Civil Society." *Polity* 23, no. 4 (1991): 505–25.

Robinson, H. M. "Prime Matter in Aristotle." *Phronesis* 19, no. 2 (January 1, 1974): 168–88.

Sacksteder, William. "Geometrical Objects." *Philosophy of Science* 48, no. 4 (2009): 573–90.

———. "Some Ways of Doing Language Philosophy: Nominalism, Hobbes, and the Linguistic Turn." *Review of Metaphysics* 34, no. 3 (1981): 459–85.

———. "Three Diverse Sciences in Hobbes: First Philosophy, Geometry, and Physics." *Review of Metaphysics* 45, no. 4 (1992): 739–72.

Saxonhouse, Arlene W. "Hobbes And The '*Horae Subsecivae*.'" *Polity* 13, no. 4 (1981): 541–67.

Schmitt, Carl. *Politische Theologie. Vier Kapitel zur Lehre von der Souveränität*. Auflage: 9, Aufl. Berlin: Duncker and Humblot, 2009.

Schrijvers, Joeri. "Anarchistic Tendencies in Continental Philosophy: Reiner Schürmann and the Hubris of Philosophy." *Research in Phenomenology* 37, no. 3 (2007): 417–39. doi:10.1163/156916407X227901.

Schürmann, Reiner. *Heidegger on Being and Acting: From Principles to Anarchy*. Bloomington: Indiana University Press, 1987.

Scotus, Ioannis Duns. *Iohanni Duns Scoti Opera Omnia*. Vol. 7. Civita: Typis Polyglottis Vaticanis, 1973.

———. *Ordinatio, Liber Primus, D. 1-2*. Vol. 2. Opera Omnia. Civitas Vaticana: Typis Polyglottis Vaticanis, 1950.

———. *Ordinatio Liber Primus, d.3*. Vol. 3. Opera Omnia. Civitas Vaticana: Typis Polyglottis Vaticanis, 1954.

———. *Ordinatio Liber Primus d.26-48*. Vol. 6. Opera. Civitas Vaticana: Typis Polyglottis Vaticanis, 1963.

———. *Quaestiones in Tertium Librum Sententiarum*. Edited by Luke Wadding. Vol. 14. Opera Omnia. Paris: Ludovicus Vives, 1894.

———. *Quaestiones Super Libros Metaphysicorum Aristotelis Libri VI–IX*. Edited by R. Andrews, G. Etzkorn, G. Gál, R. Green, F. Kelley, G. Marcil, T. Noone, and R. Wood. Vol. 4. Opera Philosophica. St. Bonaventure: The Franciscan Institute, 1997.

———. *Reportata Parisiensa*. Edited by Luke Wadding. Vol. 22. Opera Omnia. Paris: Ludovicus Vives, 1894.

———. *Reportata Parisiensia*. Edited by Luke Wadding. Vol. 23. Opera Omnia. Paris: Ludovicus Vives, 1894.

Smolak, Maciej. "Arystotelesa Charakterystyka Zniszszalnych Substancji." *Analiza I Egzystencja* 13 (2011): 27–52.

Sorell, Tom. *Hobbes*. London: Routledge, 1991.

Strauss, Leo. *Political Philosophy of Hobbes: Its Basis and Its Genesis*. Translated by Elsa M. Sinclair. Chicago: University of Chicago Press, 1952.

Tarlton, C. D. "Rehabilitating Hobbes: Obligation, Anti-Fascism and the Myth of a 'Taylor Thesis.'" *History of Political Thought* 19, no. 3 (1998): 407–38.

Taylor, A. E. "The Ethical Doctrine of Hobbes." *Philosophy* 13, no. 52 (1938).

Tuck, Richard. "Hobbes' Moral Philosophy." In *The Cambridge Companion to Hobbes*, 175–207. Cambridge: Cambridge University Press, 1996.

Von Brockdorff, Baron Cay. "Die Urform der 'Computatio Sive Logica' des Hobbes." *Journal of Philosophy* 31, no. 13 (1934): 357–58.

Warrender, Howard. *The Political Philosophy of Hobbes, His Theory of Obligation*. Oxford: Clarendon Press, 1957.

Watkins, J. W. N. "Philosophy and Politics in Hobbes." *The Philosophical Quarterly* 5, no. 19 (1955): 125–46.

Watson, George. "Hobbes and the Metaphysical Conceit." *Journal of the History of Ideas* 16, no. 4 (1955): 558–62.

Weinberger, J. "Hobbes's Doctrine of Method." *American Political Science Review* 69, no. 4 (1975): 1336–53.

Weiss, Ulrich. "Wissenschaft als meschliches Handeln. Zu Thomas Hobbes' Anthropologischer Fundierung von Wissenschaft." *Zeitschrift Fuer Philosophische Forschung* 37, no. 1 (1983): 37–55.

Yu, Jiyuan. *The Structure of Being in Aristotle's Metaphysics*. Berlin: Springer, 2003.

Zagorin, Perez. "Hobbes's Early Philosophical Development." *Journal of the History of Ideas* 54, no. 3 (1993): 505–18.

Zarka, Yves Charles. *La Décision Métaphysique de Hobbes*. Paris: J. Vrin, 1999.

———. "Liberty, Necessity and Chance: Hobbes's General Theory of Events." *British Journal for the History of Philosophy* 9, no. 3 (2001): 425–37.

Zeller, Eduard. *Die Philosophie der Griechen in ihrer geschichtlichen Entwicklung*. Hildesheim: G. Olms, 1963.

Index

www.ingramcontent.com/pod-product-compliance
Lightning Source LLC
LaVergne TN
LVHW101443200926
848483LV00001B/41
9781783486427